Fishing Entity Enforcement in High Seas Fisheries

By

Ying-Ting Chen

CAMBRIDGE
SCHOLARS
PUBLISHING

Fishing Entity Enforcement in High Seas Fisheries,
by Ying-Ting Chen

This book first published 2014

Cambridge Scholars Publishing

12 Back Chapman Street, Newcastle upon Tyne, NE6 2XX, UK

British Library Cataloguing in Publication Data
A catalogue record for this book is available from the British Library

ISBN (10): 1-4438-6388-2, ISBN (13): 978-1-4438-6388-9

For my Family

TABLE OF CONTENTS

Part II: Enforcement in High Seas Fisheries by Fishing Entities

Part III: Practice Within Regional Fisheries Organisations—The Example of Taiwan

ACKNOWLEDGEMENTS

Firstly, I would like to express my great appreciation to my supervisor, Dr Andrew L. Serdy. During the process of writing, he always provided very helpful suggestions for my thesis. Without him, this thesis will be not finished smoothly.

In addition, I am deeply grateful to my family. Their support backs me up to do the research and strive for the PhD degree.

Finally, I must offer my heartfelt thanks to those persons who did me a favor and encouraged me to finish this thesis.

INTRODUCTION

The fishery resources in the oceans used to be viewed as inexhaustible. As fishing technology has developed rapidly, however, people have gradually made efforts to adopt fishery resources reservation and management methods, rather than ways to catch as many fish as possible. Among the many aspects of fishery resources conservation and management, straddling and highly migratory stocks especially need to be protected as they stay in different areas during different stages of life and could easily face the risk of extinction.[1]

Within exclusive economic zones, coastal states can make laws to protect resources. As a part of customary law, however, the freedom of high seas results in flag state jurisdiction on the high seas. As global concern for the importance of conserving straddling and highly migratory stocks has risen, regulations concerning enforcement in the high seas to preserve these stocks and decrease their possibility of extinction have been created. To protect straddling and highly migratory stocks in various high-sea areas, the 1995 Agreement for the Implementation of the Provisions of the United Nations Convention on the Law of the Sea of 10 December 1982 Relating to the Conservation and Management of Straddling Fish Stocks and Highly Migratory Fish Stocks (UNFSA)[2] requires state parties to establish regional fishery management organisations (RFMOs) to conserve and manage those stocks. Through RFMOs, many conservation and management measures (CMMs) are established. For example, the primary method is to calculate the total allowable catch (TAC) of each stock and allocate quotas to members of RFMOs according to the TAC. Vessels might also be required to carry vessel monitoring systems (VMS) which use global positioning systems (GPS) to acquire the position of the vessels. The RFMOs then can conduct high seas enforcement, *i.e.*

---

[1] Annex I of the 1982 United Nations Convention on the Law of the Sea (UNCLOS) lists seventeen highly migratory species: albacore tuna, bluefin tuna, bigeye tuna, skipjack tuna, yellowfin tuna, blackfin tuna, little tuna, southern bluefin tuna, frigate mackerel, pomfrets, marlins, sailfish, swordfish, saury, dolphin, oceanic sharks, and cetaceans. UNCLOS was adopted on 10 December 1982 and entered into force 16 November 1994. For the full text of UNCLOS, *see A/CONF.62/122*, 1833 UNTS 397.

[2] Oceans and Law of the Sea, *A/CONF.164/37*, 2167 UNTS 3.

boarding and inspection, to determine if the vessels violate conservation and management measures.

Enforcement includes both negative and positive actions. The former means to arrest or clamp down on the violator, *i.e.* impose punishments after the law is broken; the latter, on the other hand, refers to preventing violation behaviours, *i.e.* taking proactive measures to prohibit a breach of law. Traditionally, flag state jurisdiction applies to most circumstances, except for piracy, illegal drugs trafficking, illegal broadcasting, slavery, etc.[3] Due to the international community's increased concern for fishery conservation and management, non-flag state enforcement in high seas fisheries has been extended from bilateral treaties to regional treaties. Through the management of RFMOs, members can not only exert their flag state jurisdiction but may also board and inspect each other's vessels in order to ensure compliance with related conservation and management measures.

In addition, Taiwan, officially the Republic of China, is a democratic state in East Asia neighboured by the People's Republic of China to the west, Japan to the northeast and the Philippines to the south. Taiwan's total land area is approximately 14,400 square miles, and its population 23 million. It was the world's 17th largest exporter and 18th largest importer of merchandise in 2011 and ranked the 13th in the World Economic Forum's Global Competitiveness Report 2011–2012.[4] However, due to political conflict with China, Taiwan has been denied participation in most international governmental organisations using the identity of a state or its official name. Therefore, Taiwan seeks to overcome this obstacle by using other identities or names in international forums.[5] "Fishing entity" is an identity that Taiwan uses in many RFMOs.

Although not much literature has discussed the term "fishing entity," some articles have focused on the topic of fishing entity. Andrew Serdy discussed the details of Taiwan's entry into the Commission for the Conservation of Southern Bluefin Tuna (CCSBT)[6], including Taiwan's

---

[3] See Article 110 of UNCLOS.

[4] More information about Taiwan can be found on the official websites of Taiwan's Ministry of Foreign Affairs <*http://taiwanindepth.tw* (visited on 03/02/2014) and Tourism Bureau *http://eng.taiwan.net.tw/m1.aspx?sNO=0000202* (visited on 03/02/2014).

[5] Regarding the origins of the political conflict between Taiwan and China and Taiwan's participation in international governmental organisations using different nomenclature, please *see* Chapter 7 of this book.

[6] The CCSBT was established in 1994 with the aim to ensure, through appropriate management, the conservation and optimum utilisation of the global southern

view on its position in the CCBST, and precisely analysed the CCSBT 2001 Resolution, which established an extended commission and a scientific committee to introduce the concept of the fishing entity to the CCBST.[7] Serdy also examined Taiwan's status and different position than China during process of negotiating, drafting and forming the CCBST, International Commission for the Conservation of Atlantic Tunas (ICCAT)[8], Indian Ocean Tuna Commission (IOTC)[9], Western and Central

---

bluefin tuna stock. The CCSBT is headquartered in Canberra, Australia; has five members (Australia, Japan, New Zealand, Korea and Indonesia); a fishing entity, (Taiwan); and three co-operating non-parties (the Philippines, South Africa and the European Union). Complete information on CCSBT can be found at *http://www.ccsbt.org/docs/about.html* (visited on 16/07/2010).

[7] Andrew Serdy, "Bringing Taiwan into the International Fisheries Fold: The Legal Personality of a Fishing Entity," in James Crawford and Vaughan Lowe, eds., *The British Year Book of International Law 2004* (Oxford: Clarendon Press, 2005 Vol. 75), pp. 183–221.

[8] The ICCAT is an inter-governmental fishery organisation founded in 1969 which is responsible for the conservation of tunas and tuna-like species in the Atlantic Ocean and its adjacent seas. It has 47 members: Albania, Algeria, Angola, Barbados, Belize, Brazil, Canada, Cape Verde, China, Sierra Leone, Côte d'Ivoire, Egypt, Equatorial Guinea, the European Union, France, Gabon, Ghana, Guatemala, Guinea, Honduras, Iceland, Japan, Libya, Morocco, Mauritania, Mexico, Namibia, Nicaragua, Nigeria, Norway, Panama, the Philippines, Republic of Korea, Russian Federation, Saint Vincent/Grenadines, Sao Tome and Principe, Senegal, Sierra Leone, South Africa, Syrian Arab Republic, Trinidad and Tobago, Tunisia, Turkey, the United Kingdom, United States of America, Uruguay, Vanuatu and Venezuela. In addition, the ICCAT has five co-operators: Bolivia, Chinese Taipei, Curacao, Suriname and El Salvador. Complete information can be found on the official ICCAT website <http://www.iccat.int/en/introduction.htm> (visited on 03/02/2014).

[9] The IOTC was set up in 1997 as an intergovernmental organisation mandated to manage tuna and tuna-like species in the Indian Ocean and adjacent seas. Its objective is to promote cooperation among its members in order to ensure the conservation and optimum utilisation of stocks and to encourage sustainable development of fisheries based on such stocks. Its members are Australia, Belize, China, Comoros, Eritrea, the European Community, France, Guinea, India, Indonesia, Iran, Japan, Kenya, Korea, Madagascar, Malaysia, Maldives, Mauritius, Mozambique, Oman, Pakistan, Philippines, Seychelles, Sierra Leone, Sri Lanka, Sudan, Tanzania, Thailand, the United Kingdom, Vanuatu and Yemen; Senegal and South Africa are cooperating non-contracting parties. Complete information can be found on the official IOTC website *http://www.iotc.org/English/about.php*> (visited on 04/02/2014).

Pacific Fisheries Commission (WCPFC)[10] and South East Atlantic Fisheries Organization (SEAFO).[11] Serdy focused on the subject of Taiwan, rather than "fishing entity," which seemed to be a consequence of Taiwan's participation in the RFMOs. It can be argued that the term "fishing entity" was coined mainly to refer to Taiwan, so addressing Taiwan cannot be avoided, particularly when discussing Taiwan's participation in RFMOs. In "The Emergence of the Concept of Fishing Entities: A Note," Hasjim Djalal noted that the concept of fishing entities first appeared in the UNFSA in order, he argued, to deal with the fishing vessels of Taiwan.[12] Djalal apparently regarded "fishing entity" as equivalent to Taiwan. He saw Taiwan as acting as a subject under international law and contended that, "under emerging international law, a fishing entity has also gradually become a subject of international law having the rights, obligations, and legal capacity similar to other subjects under modern international law."[13]

While considering the issue of whether fishing entities have international legal personality, Martin Tsamenyi took the positive view. He did not emphasise the links between Taiwan and fishing entities but

---

[10] The WCPFC was established in 2004 to address problems in the management of high seas fisheries resulting from unregulated fishing, over-capitalisation, excessive fleet capacity, vessel re-flagging to escape controls, insufficiently selective gear, unreliable databases and insufficient multilateral cooperation in the conservation and management of highly migratory fish stocks. The WCPFC's members are Australia, China, Canada, Cook Islands, the European Union, Federated States of Micronesia, Fiji, France, Japan, Kiribati, Republic of Korea, Republic of Marshall Islands, Nauru, New Zealand, Niue, Palau, Papua New Guinea, Philippines, Samoa, Solomon Islands, Chinese Taipei, Tonga, Tuvalu, the United States of America and Vanuatu. American Samoa, the Commonwealth of the Northern Mariana Islands, French Polynesia, Guam, New Caledonia, Tokelau and Wallis and Futuna are participating territories. Belize, the Democratic People's Republic of Korea, Ecuador, El Salvador, Indonesia, Mexico, Senegal, St Kitts and Nevis, Panama, Thailand and Vietnam are cooperating non-members. Complete information can be found on the official WCPFC website
http://www.wcpfc.int/about-wcpfc (visited on 03/02/2014).

[11] *See* Serdy, *supra note* 7, pp. 200-216. The SEAFO was founded in 2003 with the objective to ensure the long-term conservation and sustainable use of the fishery resources in SEAFO's area of competence. Its members are Angola, the European Union, Japan, Namibia, Norway, Republic of Korea and South Africa. Complete information can be found on the official SEAFO website
http://www.seafo.org/index.html (visited on 04/02/2014).

[12] Hasjim Djalal, "The Emergence of the Concept of Fishing Entities: A Note," *Ocean Development and International Law*, Vol. 37(2006), pp. 119.

[13] *Ibid.*, p. 120.

analysed international fisheries instruments, such as the UNFSA, Convention of the WCPFC, Antigua Convention, Cooperating Non-Party Schemes of the CCSBT, Inter-American Tropical Tuna Commission (IATTC)[14] and ICCAT.[15] He concluded "the international legal personality of fishing entities was confirmed by their recognition in international fisheries instruments and the creation of obligations for such entities."[16]

Michael W. Lodge, who served as the executive secretary to the Multilateral High Level Conference on the Conservation and Management of Highly Migratory Fish Stocks in West and Central Pacific from 1997 to 2000, recorded Taiwan's participation in the conference that drafted the Convention of the WCPFC. His article was entitled "The Practice of Fishing Entities in Regional Fisheries Management Organizations: The Case of the Commission for the Conservation and Management of Highly Migratory Fish Stocks in the Western and Central Pacific Ocean," but he directly used the term "Taiwan," rather than "fishing entities," throughout the article.[17] Lodge clearly considered fishing entities as equivalent to Taiwan, although he described Taiwan as a major fishing entity.[18]

Nien-Tsu Alfred Hu discussed the concept of fishing entities from Taiwan's perspective.[19] He analysed Taiwan's agreement to regard itself as

---

[14] The IATTC is the first tuna regional fisheries management organisation, set up by United States and Costa Rica in 1950 in order to conserve and manage tuna and other marine resources in the eastern Pacific Ocean and enhance scientific research and cooperation on these resources. Its members are Belize, Canada, China, Chinese Taipei, Colombia, Costa Rica, Ecuador, El Salvador, the European Union, France, Guatemala, Japan, Korea, Mexico, Nicaragua, Panama, Peru, the United States, Vanuatu and Venezuela. The Cook Islands and Kiribati are cooperating non-parties. Complete information can be found on the official IATTC website *http://www.iattc.org/HomeENG.htm* (visited on 07/08/2010).

[15] Tsamenyi had stated that only Taiwan was considered a fishing entity but did not deal with the political problems concerning Taiwan's status in international law. Instead, he focused on the legal status and content of a fishing entity. *See* Martin Tsamenyi, "The Legal Substance and Status of Fishing Entities in International Law: A Note," *Ocean Development and International Law*, Vol. 37(2006), pp. 123–131.

[16] *Ibid.*, p. 130.

[17] *See* Michael W. Lodge, "The Practice of Fishing Entities in Regional Fisheries Management Organizations: The Case of the Commission for the Conservation and Management of Highly Migratory Fish Stocks in the Western and Central Pacific Ocean," *Ocean Development and International Law*, Vol. 37(2006), pp. 185–207.

[18] *Ibid.*, p. 200.

[19] Nien-Tsu Alfred Hu, "Fishing Entities: Their Emergence, Evolution, and Practice from Taiwan's Perspective," *Ocean Development and International Law*, Vol. 37(2006), pp. 149–183.

a fishing entity as set out in the UNFSA and Taiwan's use of the identity of fishing entity to participate in the International Scientific Committee for Tuna and Tuna-like Species in the North Pacific Ocean (ISC)[20], IATTC and WCPFC.[21] Hu stated that a fishing entity is a subject of international law and should be regarded as "an entity possessing full autonomy in the conduct of its external fisheries relations and of all matters provided for in relevant international law."[22] In addition, Peter S. C. Ho discussed the impact of the UNFSA on Taiwan's participation in RFMOs.[23] He observed that, before the adoption of the UNFSA, Taiwan did not belong to any RFMO, but the creation of the term "fishing entity" by the UNFSA allowed Taiwan to join some RFMOs and become further involved in their decision making.[24] Similarly to Hu, Ho also discussed Taiwan's decision to be regarded as a fishing entity and the process through which Taiwan participated in some RFMOs, including the ICCAT, WCPFC, CCSBT and IATTC.[25]

Dustin Kuan-Hsiung Wang examined Taiwan's role during the drafting of the Antigua Convention, including the application of the term "fishing entities" in the convention, Taiwan's viewpoint in each meeting, other states' views of Taiwan's position in the convention and the conflicts between Taiwan and China.[26] Furthermore, William Edeson in his article

---

[20] The United States and Japan founded the ISC in 1995 with the objective to improve scientific research and cooperation in the conservation and rational utilisation of the species of tuna and tuna-like fishes which inhabit the North Pacific Ocean during part or all of their life cycle. The ISC was also charged with laying the scientific groundwork, if at some point in the future, it is decided to create a multilateral regime for the conservation and rational utilisation of these species in this region. ISC members are: Canada, Chinese Taipei, Japan, Republic of Korea, Mexico, People's Republic of China and the United States of America. Observers include the IATTC, FAO, North Pacific Marine Science Organization (PICES), Secretariat of the Pacific Community (SPC) and WCPFC. Further information can be found on the official ISC website http://isc.ac.affrc.go.jp/ (visited on 05/08/2010).

[21] Hu, *supra note* 19.

[22] *Ibid.*, p. 175.

[23] Peter S.C. Ho, "The Impact of the Fish Stocks Agreement on Taiwan's Participation in International Fisheries Fora," *Ocean Development and International Law*, Vol. 37(2006), pp. 133–148.

[24] *Ibid.*

[25] *Ibid.*

[26] Dustin Kung-Hsiung Wang, "Taiwan's Participation in Regional Fisheries Management Organizations and the Conceptual Revolution on Fishing Entity: The

"Some Future Directions for Fishing Entities in Certain Regional Fisheries Management Bodies" identified two types of RFMOs: those outside the Food and Agriculture Organization (FAO) of the United Nations (UN) framework and those within the UN context.[27] Regarding the first type, Edeson roughly described the RFMOs which have regulations about fishing entities, including the CCBST, IATTC and WCPFC.[28] He suggested that the ICCAT, in particular, consider the experience of the CCSBT in establishing an extended commission and scientific committee to ensure that "Taiwan gained benefits from its participation in the fisheries covered by ICCAT commensurate with its commitment to comply with applicable conservation and management measure."[29] Edeson mainly focused on the second type of RFMOs, specifically the IOTC, which he pointed out was so far the only RFMO concerned with tuna which ruled out the participation of fishing entities.[30] He analysed the problem facing the IOTC and suggested that the practical option to position the IOTC outside the FAO framework would be to adopt amendments to the IOTC Agreement without creating new obligations for the contracting parties.[31] In another article, "An International Legal Extravaganza in the Indian Ocean: Placing the Indian Ocean Tuna Commission outside the Framework of FAO," Edeson focused more about the legal obstacles placed in the way of this process by the FAO's internal legal advisers.[32]

The only fishing entity at present is Taiwan, so the discussed research on fishing entities focused mainly on Taiwan's participation in various tuna RFMOs. Most authors attended the relevant meetings of the RFMOs, so they could provide first-hand records of the meetings and share their perspectives of the negotiation process, resulting in significant research on fishing entities. Thus, most of the literature discussed Taiwan's perspective

---

Case of IATTC," *Ocean Development and International Law*, Vol. 37(2006), pp. 209–219.

[27] William Edeson, "Some Future Directions for Fishing Entities in Certain Regional Fisheries Management Bodies," *Ocean Development and International Law*, Vol. 37(2006), pp. 245–264.

[28] *Ibid.*, pp. 248–251.

[29] *Ibid.*, pp. 248–250.

[30] *Ibid.*, pp. 251–261.

[31] *Ibid.*

[32] W.R. Edeson, "An International Legal Extravaganza in the Indian Ocean: Placing the Indian Ocean Tuna Commission outside the Framework of FAO," *The International Journal of Marine and Coastal Law*, Vol. 22, No.4(2007), pp. 485-515.

on fishing entities, rather than fishing entities themselves. Although the term "fishing entities" was coined primarily to resolve the difficulty of Taiwan's participation in relevant fisheries organisations, the term itself needs to be further discussed as it is included in international fisheries law. Therefore, this book will focus on the topic of fishing entities as a new concept in international fisheries law. It is impossible to avoid mentioning Taiwan while discussing fishing entities; however, this book will concentrate on fishing entities themselves and treat Taiwan as an example, not as a synonym.

In addition, a fishing entity, as an actor in the international law of the sea, might, like Taiwan, possess advanced technology in fishing skills, so it cannot be ignored in global and regional conservation and management of fishery resources. Although there is not a legal or normal definition of fishing entity, a fishing entity is definitely categorised as an "entity" rather than a "state," resulting in certain unclear circumstances for its involvement in global or regional agreements whose subjects are assumed to be states. This ambiguity is especially strong in high seas enforcement actions concerning the jurisdiction of a state, such as boarding and inspection.

However, the concept of conservation and management should be based on the premise that all actors must be brought under regulations; otherwise, the actors' efforts within regulations would be in vain and lead to failure. Therefore, it is necessary and important to study the enforcement of fishing entities on the high seas in order to ensure the effectiveness of the rules conserving and managing the straddling and highly migratory stocks. Since the WCPFC became the first RFMO to adopt its own boarding and inspection procedure, no relevant literature discussing fishing entities' position in high seas enforcement has been produced. This book thus stands as the first research to consider together fishing entities and high seas enforcement in the international law of the sea.

In this book, I link fishing entities to high seas enforcement by discussing the role of fishing entities in international law of the sea and clarifying their obligations and rights in high seas fishery enforcement, especially regarding regulations concerning conservation and management of straddling and highly migratory stocks. This book also discusses RFMOs which focus on the protection of straddling and highly migratory stocks and allow participation by fishing entities. Before moving to the linkage between fishing entities and high seas fisheries enforcement, it is necessary to review the concept of fishing entities, including the international instruments in which the term "fishing entities" originates.

Thus, the first question this book attempts to answer is: (a) Why is the issue of fishing entities important? What is the legal status of the fishing entity in the international law of the sea and in RFMOs?

After introducing the concept of fishing entities, the book further considers the position of fishing entities in fisheries enforcement on the high seas and seeks to answer the following research questions: (b) How does fisheries enforcement on the high seas treat the fishing entity which possesses two disparate identities—a flag state and a non-flag state? Based on the role that fishing entities should play, the next question asked is: (c) What problems might fishing entities and the international community face under present fishery regulations? The regulations of fisheries enforcement on the high seas are mostly set out in the UNFSA, which requires RFMOs to adopt their own boarding and inspection procedures; therefore, while understanding the concept of the fishing entity and its status in international fisheries enforcement on the high seas, the book examines the practices of high seas fisheries enforcement performed by fishing entities within RFMOs. Thus arises the question: (d) What practices for fishing entity enforcement on high seas do RFMOs adopt?

To answer these questions, the book is divided into three parts: the fishing entity, the link between the fishing entity and its fisheries enforcement on the high seas, and the practice of fishing entity and of others regarding it in RFMOs. In Part 1 of this book entitled "The Existence of the Fishing Entity," I first introduce the concept of the fishing entity in the first chapter. As the fishing entity exists, it leads us to consider its legal status in the international law of the sea, which is the topic of Chapter 2. While discussing the fishing entity's status in international law, especially under the UNFSA, it is found that the fishing entity primarily plays a role in many individual RFMOs. Therefore, Chapter 3 "Fishing Entities in RFMOs" addresses the regulations concerning fishing entities in the main eight RFMOs governing most high sea areas regarding the conservation and management of straddling and highly migratory fish stocks.

In the second part of this book entitled "Enforcement in High Seas Fisheries by Fishing Entities," it first is shown that a state's enforcement on the high seas can be categorised as enforcement against its own vessels, which falls under flag state jurisdiction, and as enforcement against other state's vessels, which is called non-flag state enforcement. I explore these two concepts in Chapters 4 and 5, specifically whether fishing entities can exercise flag state jurisdiction and non-flag state enforcement. Although a fishing entity might be similar to a state in flag state and non-flag state enforcement, it is not a state after all and moreover is a new concept in the international law of the sea. Thus, it might encounter some problems in this field, which are addressed in Chapter 6, "The Problematic Consequences of Fishing Entity Enforcement on the High Seas."

In the third part of this book entitled "Practice with RFMOs—The Example of Taiwan," I first introduce the background of and reasons why Taiwan became a fishing entity and how it participated as a members in three RFMOs (WCPFC, IATTC and ISC), as discussed in Chapter 7. Chapter 8 continues to examine Taiwan's participation in three other RFMOs (CCSBT, ICCAT and IOTC) as a non-member. I do not discuss the North Pacific Anadromous Fish Commission (NPAFC)[33] or the SEAFO because Taiwan does not participate in the SEAFO at all, and although it has occasionally attended meetings of the NPAFC as an observer since 2005, Taiwan's involvement in this organisation is different than in the ICCAT. Taiwan is not bound by the NPAFC's regulations but only provides scientific information to it. Most importantly, while attending NPAFC meetings, Taiwan does not use the identity of a fishing entity but the name "Taiwan" and is not introduced or recorded as a fishing entity.[34]

---

[33] The NPAFC was established in 1993 with the objective to promote the conservation of anadromous stocks in the convention area. The contracting parties are Canada, Japan, Republic of Korea, the Russian Federation and the United States. Complete information can be found on the official NPAFC site *http://www.npafc.org/new/about_convention.html* (visited on 07/02/2014).

[34] *See* the 2005–2010 and 2012 annual reports of the NPAFC, available at *http://www.npafc.org/new/pub_annualreport.html* (visited on 07/02/2014).

LIST OF ABBREVIATIONS

APEC	Asia–Pacific Economic Cooperation
CCLM	Committee on Constitutional and Legal Matters
CCSBT	Commission for the Conservation of Southern Bluefin Tuna
CIESIN	Center for International Earth Science Information Network
CITES	Convention on International Trade in Endangered Species
CMM	Conservation and management measure
COFI	Committee on Fisheries
EEC	European Economic Community
EEZ	Exclusive economic zone
FAO	Food and Agriculture Organization of the United Nations
FFA	Pacific Islands Forum Fisheries Agency
GPS	Global positioning system
KMT	Kuo Min Tang party
IATTC	Inter-American Tropical Tuna Commission
ICA	International Council for Information Technology in Government Administration
ICCAT	International Commission for the Conservation of Atlantic Tunas
ICN	International Competition Network
ICNAF	International Commission for the Northwest Atlantic Fisheries
ICJ	International Court of Justice
INPFC	International North Pacific Fisheries Commission
IOC	International Olympic Committee
IOTC	Indian Ocean Tuna Commission
IPOA	International plan of action
IUU	Illegal, unregulated and unreported fishing
MCS	Monitoring, control and surveillance
MHLC	Multilateral High-Level Conference in the Conservation and Management of Highly Migratory Fish Stocks in the Western and Central Pacific
NAFO	Northwest Atlantic Fisheries Organization
NGOs	Non-governmental organisations

NPAFC	North Pacific Anadromous Fish Commission
PICES	North Pacific Marine Science Organization
PLO	Palestine Liberation Organization
PRC	People's Republic of China
PSI	Proliferation Security Initiative
PWG	Permanent Working Group for the Improvement of ICCAT Statistics and Conservation Measures
RFMO	Regional fisheries management organization
ROC	Republic of China
SBT	Southern bluefin tuna
SEAFO	South East Atlantic Fisheries Organisation
SIOFA	South Indian Ocean Fisheries Agreement
SPC	Secretariat of the Pacific Community
TAC	Total allowable catch
TCC	Technical and compliance committee
UN	United Nations
UNCLOS	United Nations Convention on the Law of the Sea
UNFSA/ UNIA	Agreement for the Implementation of the Provisions of the United Nations Convention on the Law of the Sea of 10 December 1982 Relating to the Conservation and Management of Straddling Fish Stocks and Highly Migratory Fish Stocks
US	United States
VMS	Vessel monitoring system
WCPFC	Western and Central Pacific Fisheries Commission
WMDs	Weapons of mass destruction
WTO	World Trade Organization

PART I:

FISHING ENTITY

CHAPTER ONE

THE EXISTENCE OF THE FISHING ENTITY

1. Introduction

After decades of efforts to establish an effective regime to maintain the sustainability of straddling and highly migratory fish stocks, the Fish Stock Agreement was adopted on 4 August 1995 and entered into force on 11 December 2001 after deposit of the thirtieth instrument of accession by the Republic of Malta. Due to the necessity of cooperating in and the indivisibility of the fields of fisheries conservation and management, the UNFSA introduces the concept of fishing entity into its regulations.[1] Although the agreement does not give the term "fishing entity" a clear definition, it implies not only the special status of the fishing entity in the international law of the sea but also the concern that the fishing entities should not be excluded from international collaboration in conserving and managing fisheries resources.

According to FAO statistics, more than 70% of fisheries are overexploited or depleted. Highly migratory and straddling stocks are affected particularly seriously. Global fish production has continued to increase and reached 148.5 million tonnes in 2010.[2] Although catches of

---

[1] Article 1(3) states that "[t]his Agreement applies *mutatis mutandis* to other fishing entities whose vessels fish on the high seas," and article 17(3) also mentions "fishing entity:" "States which are members of a subregional or regional fisheries management organization or participants in a subregional or regional fisheries management arrangement shall, individually or jointly, request the fishing entities referred to in article 1, paragraph 3, which have fishing vessels in the relevant area to cooperate fully with such organization or arrangement in implementing the conservation and management measures it has established, with a view to having such measures applied de facto as extensively as possible to fishing activities in the relevant area. Such fishing entities shall enjoy benefits from participation in the fishery commensurate with their commitment to comply with conservation and management measures in respect of the stocks."

[2] Food and Agriculture Organization of the United Nations, *FAO Yearbook: Fishery and Aquaculture Statistics* (Rome: FAO Fisheries Department, 2012), p. xvi.

tuna and tuna-like species have decreased by 1.7 per cent and remained stable since 2010 after an upward trend which led to the historical peak catch in 2006,[3] they still belong to species urgently needing to be protected. During the 13–25 March 2010 Conference of the Parties at the 15th Meeting of the Convention on International Trade in Endangered Species of Wild Fauna and Flora (CITES)[4] in Doha (Qatar), Morocco proposed a ban on the trade of Atlantic bluefin tuna.[5] Although this proposal was not adopted, it proved useful in drawing international attention to not only the high possibility of commercial extinction faced by the bluefin tuna but also the importance of the conservation and sustainable development of fishery resources. Due to the vulnerability of the highly migratory and straddling stocks, their conservation and management is of global concern. A fishing entity would be exempt from international fishery conservation and management regimes if it were not regarded as an actor in international fishery regulations, a situation which would worsen if the entity possessed the ability to severely deplete the stocks and cause tension within the regime.

This chapter aims to introduce the concept of fishing entity and review the stipulations on fishing entities concerning fishery resources conservation and sustainable development by considering the relations between the fishing entity and the related international instruments.

2. The Concept of the Fishing Entity

States are the primary subjects in the traditional international law of the sea, as well as the main actors in international society.[6] After the two

---

[3] *Ibid.*, also *see* Food and Agriculture Organization of the United Nations, *FAO Yearbook: Fishery and Aquaculture Statistics* (Rome: FAO Fisheries Department, 2009), p. xxii.

[4] CITES is an international agreement between governments. It was drafted as a result of a resolution adopted in 1963. The text of the convention was finally agreed upon at a meeting of representatives of 80 countries in Washington, D.C. on 3 March 1973, and it entered into force on 1 July 1975. The aim of CITES is to ensure that the international trade in specimens of wild animals and plants does not threaten their survival. For further information, *see* the official CITES website, <http://www.cites.org/> (visited on 15/03/2010).

[5] *See* CITES, *http://www.cites.org/eng/cop/15/sum/E15-Com-I-Rec08.pdf* (visited on 30/09/2010).

[6] In the theory of international relations, the neo-realist emphasises the role of state as an actor, whereas the idealist focuses on the important roles of other actors, such as international organisations, transnational corporations and individuals, etc.

World Wars, actors other than states have emerged as subjects of international law have emerged, such as international organisations, international non-government organisations, and entities.[7] It is thought that an entity with limited rights and obligations and limited capacity to make an international claim can be regarded as a legal person.[8] However, an entity which does not satisfy those conditions might still have a legal personality generated by certain international agreements.[9] The most prominent example is the Holy See, whose exclusive sovereignty and jurisdiction over the City of the Vatican was recognised in the international domain by Italy in the 1929 Treaty and Concordat;[10] afterward, it was recognised by most states as well.[11]

In addition to this religious entity, non-self-governing territories are another well-known entity. The mandate territories under the League of Nations and the trust territories of the UN are regarded as political entities. National liberation movements can also be recognised as political entities, for example, the Palestine Liberation Organization (PLO) was granted observer status in the UN General Assembly.[12] Similar to national liberation movements, insurgent communities *de facto* occupy and control a specific territory during a civil war within a country and can be recognised as belligerents, which possess a certain international personality.[13]

In addition to religious and political entities, the concept of entity extends to the economic sphere. All members of the Asia–Pacific Economic Cooperation (APEC) are called "economies," which are economic entities.[14] Parallel to its status as an economic entity, Taiwan became a member of the World Trade Organization (WTO) under the

---

However, both claim that states are the primary, not the only actors, in international relations.

[7] Hasjim Djalal, "The Emergence of the Concept of Fishing Entities: A Note," *Ocean Development and International Law*, Vol. 37(2006), p. 117.

[8] Ian Brownlie, *Principles of Public International Law*, 6th ed. (New York: Oxford University Press, 2003), p. 57.

[9] *Ibid.*

[10] *Ibid.*, p. 63.

[11] *See* Robert A. Graham, *Vatican Diplomacy* (Princeton: Princeton University Press, 1959), pp. 19–21. However, it now recognised as a state, instead of an entity.

[12] Brownlie, *supra note* 8, pp. 61–62.

[13] *Ibid.*, p. 63.

[14] APEC, *http://www.apec.org/apec/member_economies.html* (visited on 19/02/2010).

identity of customs entity using the name "the Separate Customs Territory of Taiwan, Penghu, Kinmen and Matsu" in 2002.[15]

Another type of entity characterised by its economic functions is the fishing entity. The 1995 Fish Stock Agreement marked the first time that the concept of fishing entity was written into the provisions of a formal international agreement. However, the agreement does not definitely define a fishing entity's legal status and its specific rights and obligations in international law. Further regulation directly related to fishing entities can be seen in the 1995 Code of Conduct for Responsible Fisheries.[16] Article 1(2) states that "[t]he Code is global in scope, and is directed toward members and non-members of FAO, fishing entities, subregional, regional and global organizations," and Article 4(1) that "[a]ll members and non-members of FAO, fishing entities and relevant subregional, regional and global organizations … management and utilization of fisheries resources and trade in fish and fishery products should collaborate in the fulfilment and implementation of the objectives and principles contained in this Code."[17]

Provisions related to fishing entities can usually be seen in RFMOs as well. To encourage entities with vessels fishing for southern bluefin tuna to implement the CCSBT's conservation and management measures, the commission's members tried to settle the difficulty of Taiwan's participation as a member in CCSBT, leading to the Resolution to Establish an Extended Commission and an Extended Scientific Committee adopted in April 2001.[18] Taiwan, named "Fishing Entity of

---

[15] *See* Andrew Serdy, "Bringing Taiwan into the International Fisheries Fold: The Legal Personality of a Fishing Entity," in James Crawford and Vaughan Lowe, eds., *The British Year Book of International Law 2004* (Oxford: Clarendon Press, 2005, Vol. 75), pp. 217–218. Also *see* WTO, <*http://www.wto.org/english/thewto_e/countries_e/chinese_taipei_e.htm* (visited on 19/02/2010).

[16] To ensure the sustainable development of global fisheries, the FAO in 1995 adopted the Code of Conduct for Responsible Fisheries, which regulates the principles and international standards of behaviour for responsible practices in order to ensure the effective conservation, management and development of living aquatic resources with respect for the ecosystem and biodiversity. It is a non-binding instrument but contains a large number of regulations in 12 articles.

[17] *See* FAO, *http://www.fao.org/DOCREP/005/v9878e/v9878e00.htm* (visited on 22/02/2010).

[18] *See* Serdy, *supra note* 15, pp. 184–199. The full text of the resolution can be seen on the CCSBT website, *http://www.ccsbt.org/docs/pdf/about_the_commission/the_Extended_commission.pdf* (visited on 16/08/2010).

Taiwan," became a member of the Extended Commission in 2002 through an exchange of letters.[19]

In addition to the CCSBT, the Charter of the ISC provides that coastal states, other states, and fishing entities in the region, or with vessels fishing for these species in the region are eligible to become members of ISC. Pursuant to the charter, Taiwan, as a fishing entity, became a member under the name of "Chinese Taipei" in 2002.[20]

Furthermore, in July 2003, the IATTC replaced the 1949 IATTC Convention with the Convention for the Strengthening of the Inter-American Tropical Tuna Commission Established by the 1949 Convention between the United States of America and the Republic of Costa Rica (hereinafter the Antigua Convention),[21] with stipulations concerning fishing entities as well. Article XIX states that "Article XVIII of this Convention applies, *mutatis mutandis*, to fishing entities that are members of the Commission," while Article XXI stipulates that the responsibilities of fishing entities as members of the commission are the same as those of other flag states outlined in Article XX. Article XVIII contains further provisions directly applicable to fishing entities. Previously an observer, Taiwan became a member as a fishing entity with the name of "Chinese Taipei" when the Antigua Convention entered into force on 27 August 2010.

In addition to the Antigua Convention, Annex I of the Convention on the Conservation and Management of Highly Migratory Fish Stocks in the Western and Central Pacific Ocean[22] has particular stipulations for fishing entities. Article 9(2) states that "[a] fishing entity referred to in the Agreement, which has agreed to be bound by … this Convention in accordance with the provisions of Annex I, may participate in the work … of the Commission." Annex I of the convention further clarifies fishing entities' status in the IATTC. Firstly, any fishing entity whose vessels fish

---

[19] Peter S.C. Ho, "The Impact of the Fish Stocks Agreement on Taiwan's Participation in International Fisheries Fora," *Ocean Development and International Law*, Vol. 37(2006), p. 141.

[20] The full text of the charter can be seen on the official ISC website, *http://isc.ac.affrc.go.jp/about_isc/charter.html* (visited on 05/08/2010).

[21] *See* IATTC, *http://www.iattc.org/PDFFiles2/Antigua_Convention_Jun_2003.pdf* (visited on 05/08/2010).

[22] The full text of the convention is available at *http://www.wcpfc.int/doc/convention-conservation-and-management-highly-migratory-fish-stocks-western-and-central-pacific-* (visited on 22/08/2010). This convention established the WCPFC, which was the first regional fisheries agreement to be drafted after the adoption of the 1995 Fish Stock Agreement.

for highly migratory fish stocks in the convention area has the right to agree to be bound by the convention and withdraw from such agreement. Secondly, fishing entities shall participate in the work of the commission and comply with the obligations in this convention. References to the IATTC or its members include, for the purposes of the convention, such fishing entities as well as contracting parties. In addition, the Permanent Court of Arbitration will settle any disputes concerning the interpretation or application of the convention involving a fishing entity. Under those articles, Taiwan, as a fishing entity, signed a document called "Arrangement for the Participation of Fishing Entities"[23] under the name of "Chinese Taipei."

Obviously, although most RFMOs established certain provisions related to fishing entities in order to fulfil the organisations' conservation and management measures, they do not create any definition or give any explanation of "entity" and "fishing entity" but simply use this term. The term "fishing entities" is used by most RFMOs; however, their status within RFMOs is different than that of states and varies among RFMOs.

3. International Instruments Concerning Fishing Entities

3.1 Food and Agriculture Organization

In the 1992, the FAO Technical Consultation on High Seas Fishing made the first reference to fishing entities in a paper entitled "International Fishery Bodies: Considerations for High Seas Management."[24] In this

---

[23] *See* WCPFC, *http://www.wcpfc.int/doc/arrangement-participation-fishing-entities* (visited on 22/08/2010). This arrangement states, "The Conference HEREBY INVITES Chinese Taipei, as a fishing entity, and Chinese Taipei HEREBY DECLARES its intent:

(a) to participate in the Preparatory Conference established by the resolution attached to the Final Act of the Conference,

(b) subject to the fulfilment of its domestic legal requirements, to agree to be bound by the regime established by the Convention in accordance with article 9, paragraph 2, of the Convention, and to participate in the Commission for the Conservation and Management of Highly Migratory Fish Stocks in the Western and Central Pacific Ocean in accordance with the Convention."

[24] Food and Agriculture Organization of the United Nations, *Fisheries Report No. 484 Supplement*, FIPL/R484(Suppl.), Papers presented at the FAO Technical Consultation on High Seas Fishing, Rome, 7–15 September 1992, pp. 44–54. Also *see* Jean-Pierre Lévy and Gunnar G. Schram, eds., *United Nations Conference on Straddling Fish Stocks and Highly Migratory Fish Stocks—Selected Documents* (Hague: Kluwer Law International, 1996), pp. 346–358. Also *see* Nien-Tsu Alfred

paper, the issue of non-contracting parties is emphasised in paragraph 45: "The treatment of non-contracting parties is an important and real issue that should be addressed in the context of high-seas fisheries management. Some nations or other entities operating in a fishery may opt not to participate in a high seas management body or they may be excluded from it (e.g., for political or other reasons). The effectiveness of high seas management will therefore be significantly reduced if a major entity in a fishery does not participate in determining management decisions and in turn is not bound by those decisions."[25]

Although paragraph 45 uses the phrases "entities operating in a fishery" and "a major entity in a fishery" rather than "fishing entities," paragraph 46 makes an explicit reference to "fishing entities:" "The exclusion of parties from management bodies for political or other reasons poses particular difficulties. Taiwan (Province of China) is a major international fishing entity. Its high seas fishing capacity is extensive and likely to increase, especially in the Indian and South Pacific Oceans. However, due to political non-recognition, Taiwan (Province of China) does not participate fully in any fishery management bodies. Similarly, legal constraints prevent the EEC from participating in some fishery bodies."[26]

This FAO document not only discusses the problems that the non-contracting parties and fishing entities may face but also implies the close connection between non-contracting parties and fishing entities: fishing entities are usually also non-contracting parties in RFMOs.[27] This paper also clearly points out that non-contracting parties can cause the problem of unregulated fishing, which undermines the benefits generated by conservation and management measures. This document intimates the importance and necessity of bringing fishing entities into the decision-making process of high seas management measures in order to ensure that each body will be bound by those decisions, increasing the effectiveness of the related measures.

3.2 The United Nations Fish Stock Agreement

The provisions in the Fish Stock Agreement which mention fishing entities are Articles 1(3) and 17(3). Article 1(3) states "[t]his Agreement applies

---

Hu, "Fishing Entities: Their Emergence, Evolution, and Practice from Taiwan's Perspective," *Ocean Development and International Law*, Vol. 37(2006), p. 150.

[25] Food and Agriculture Organization of the United Nations, *ibid.*, p. 52.

[26] *Ibid.*

[27] For example, Taiwan is a member of the IATTC and WCPFC as a fishing entity, but is not a contracting party of the conventions that created them.

mutatis mutandis to other fishing entities whose vessels fish on the high seas." Article 17(3) requires members of RFMOs to cooperate with fishing entities.[28] These two provisions try to cover the activities of fishing entities with different objects. Article 1(3) applies to any contracting parties of the agreement with the identity of fishing entities; therefore, all fishing entities that have acceded to or ratified this agreement would have the same rights and responsibilities as "normal" contracting parties. This provision also seems to imply that the agreement does not exclude fishing entities from becoming parties of it. Strangely, though, articles 37 and 39 concerning signature and accession, respectively, do not allow all fishing entities to participate.[29] However, the objects of Article 17(3) are the members of the RFMOs concerning straddling and highly migratory stocks. This article requires those members to cooperate with the fishing entities mentioned in Article 1(3) to enforce the organisations conservation and management measures of. In addition, under Article 17(3), fishing entities which agree to conform to the conservation and management measures for these stocks should receive the commensurate benefits of fishing entities from participation in the fishery.

3.3 1995 Code of Conduct for Responsible Fisheries

The scope of geography and objects to which the 1995 Code of Conduct for Responsible Fisheries applies are indicated by Article 1(2). This article specifies that the code is global and targeted toward members and non-members of FAO, fishing entities and government and nongovernmental subregional, regional and global organisations. As well, it applies to all persons concerned with the conservation of fishery resources and management and development of fisheries, such as fishers, those engaged in processing and marketing of fish and fishery products and other users of the aquatic environment in relation to fisheries.

In addition, Article 4(1) states that all related actors, which include fishing entities, should collaborate in the fulfilment and implementation of the code's objectives and principles concerning the conservation, management and utilisation of fisheries resources and trade in fish and fishery products. Clearly, in these two articles, fishing entities are regarded as an actor with responsibilities for their fisheries. Although the code is hortatory and has no direct binding effect in international law, that does

---

[28] For the text of Article 17(3), please *see supra note* 1.

[29] Whether the fishing entities could become parties of the agreement is discussed in Chapter 2 of this book.

not reduce the importance of fishing entities as actors of responsible fisheries.

3.4 Four International Plans of Action by the Food and Agriculture Organization

In 1998, the FAO Committee on Fisheries (COFI) found it necessary to establish certain forms of international agreement in order to achieve or implement the 1995 Code of Conduct for Responsible Fisheries. [30] Therefore, two intergovernmental meetings, open to all FAO members, were held in 1998 at which discussions resulted in three international plans of action (IPOAs).[31] Those three texts were finalised as the IPOA for the Management of Fishing Capacity, the IPOA for the Conservation and Management of Sharks and the IPOA for Reducing Incidental Catch of Seabirds in Long-line Fisheries. They were adopted at the 23rd Session of COFI in February 1999.[32]

At the same session, COFI determined that it was necessary to prevent, deter and eliminate illegal, unreported and unregulated (IUU) Fishing.[33] Next, an Expert Consultation was held in Australia from 15–19 May 2000, leading to a basic text for the subsequent FAO Technical Consultations on IUU Fishing in October 2000 and February 2001.[34] Finally, the IPOA for IUU Fishing was approved by COFI consensus on 2 March 2001 and endorsed at the 120th Session of the FAO Council on 23 June 2001.[35]

The provisions in these four IPOAs concerning fishing entities are as follows.

I. International Plan of Action for the Management of Fishing Capacity
Paragraph 5 reads: "This document is in furtherance of the commitment of all States to implement the Code of Conduct. States and regional fisheries

---

[30] Food and Agriculture Organization of the United Nations, *International Plan of Action for Reducing Incidental Catch of Seabirds in Longline Fisheries, International Plan of Action for the Conservation and Management of Sharks, International Plan of Action for the Management of Fishing Capacity* (Rome: FAO, 1999), p. iii.
[31] *Ibid.*
[32] *Ibid.*
[33] Food and Agriculture Organization of the United Nations, *International Plan of Action to Prevent, Deter and Eliminate illegal, Unreported and Unregulated Fishing* (Rome: FAO, 2001), p. iii, p. 1.
[34] *Ibid.*
[35] *Ibid.*

organizations should apply this document consistently with international law and within the framework of the respective competencies of the organizations concerned."[36] Note 7 clearly explains that, in this document, the term "state" includes members and non-members of the FAO and also applies *mutatis mutandis* to "fishing entities" other than states.[37]

II. International Plan of Action for the Conservation and Management of Sharks

Similarly, paragraph 10 of the IPOA for the Conservation and Management of Sharks states that it has been elaborated within the framework of the Code of Conduct for Responsible Fisheries.... The provisions of Article 3 of the Code of Conduct apply to the interpretation and application of this document and its relationship with other international instruments. All concerned States are encouraged to implement it.[38] Note 6 explains that, in this document, the term "state" includes members and non-members of FAO and applies *mutatis mutandis* to "fishing entities" other than states.[39]

III. International Plan of Action for Reducing Incidental Catch of Seabirds in Long-line Fisheries

Paragraph 8 of this IPOA for Reducing Incidental Catch of Seabirds in Long-line Fisheries is almost a duplicate of paragraph 10 of the IPOA for the Conservation and Management of Sharks and it reads: "IPOA-Seabirds ... has been elaborated within the framework of the Code of Conduct for Responsible Fisheries. ... The provisions of Article 3 of the Code of Conduct apply to the interpretation and application of this document and its relationship with other international instruments. All concerned States are encouraged to implement it."[40] Like the IPOA for the Management of Fishing Capacity, note 3 of the IPOA for Reducing Incidental Catch of Seabirds in Long-line Fisheries interprets that the term "state" as applying *mutatis mutandis* to "fishing entities" other than states.[41]

---

[36] Food and Agriculture Organization of the United Nations, *supra note* 30, p. 19. (emphasis added)
[37] *Ibid.*
[38] Food and Agriculture Organization of the United Nations, *supra note* 30, p. 12. (emphasis added)
[39] *Ibid.*
[40] Food and Agriculture Organization of the United Nations, *supra note* 30, p. 2. (emphasis added)
[41] *Ibid.*

IV. International Plan of Action to Prevent, Deter and Eliminate Illegal, Unreported and Unregulated Fishing

The provisions of paragraph 3(3) defining "unregulated fishing" mention fishing entities. Paragraph 3(3)(1) reads: "Unregulated fishing refers to fishing activities in the area of application of a relevant regional fisheries management organization that are conducted by … a fishing entity, in a manner that is not consistent with or contravenes the conservation and management measures of that organization."[42] That the IPOA for IUU places fishing entities under the definition of unregulated fishing, instead of illegal fishing or unreported fishing, implies the special status of fishing entities whose fishing activities might not rise to illegal or unreported fishing but have some potential to involve the problem of unregulated fishing. This potential can reasonably highlight the need to regulate the conduct of fishing entities.

In addition, paragraph 5 echoes the related provisions of the 1995 Code of Conduct for Responsible Fisheries on fishing entities. It states that "[t]he IPOA is also directed as appropriate towards fishing entities as referred to in the Code of Conduct."[43]

These four IPOAs implement the 1995 Code of Conduct for Responsible Fisheries. Although these are all hortatory, their provisions explicitly concern international society. As the existence of fishing entities cannot be denied, they are regarded as the subject of the provisions of these four IPOAs and the 1995 Code of Conduct for Responsible Fisheries. Therefore, fishing entities should be participants in related agreements and RFMOs so that their regulations can be complied with and implemented completely.

4. Conclusion

This section introduces the international instruments with provisions that directly address fishing entities. A review of these stipulations gives a picture of the general concept of the fishing entity. It can be concluded that fishing entities are considered as important as states in fishery resources conservation and sustainable development. In addition to these international instruments, still other RFMOs' conventions and guidelines bring fishing entities into their systems, including the CCSBT, Guidelines for the Interim Scientific Committee for Tuna and Tuna-like Species in the North Pacific Ocean, WCPFC Convention, Antigua Convention and

---

[42] Food and Agriculture Organization of the United Nations, *supra note* 33, p. 2.

[43] *Ibid.*, p. 3.

Convention on the Conservation and Management of Fishery Resources in the South-East Atlantic Ocean. These conventions and related documents are discussed in the third chapter.

CHAPTER TWO

THE LEGAL STATUS OF THE FISHING ENTITY

1. Introduction

The UNFSA brought the concept of fishing entities into international fisheries law. However, it neither defines fishing entities not clearly addresses their legal status, rights and obligations under international law. Given the significant role that fishing entities play in international fisheries law, it is necessary to determine if they possess international legal personality and what their legal status is under international law. Firstly, this chapter discusses what personality is in international law and the role that fishing entities play under international fisheries law. Next, it examines the legal status of fishing entities under the 1982 United Nations Convention on the Law of the Sea (UNCLOS), which is regarded as a significant convention in international fisheries law. Finally, this chapter reviews the process by which the concept of fishing entities was introduced into the UNFSA during the UN Conference on Straddling Fish Stocks and Highly Migratory Fish Stocks and analyses whether fishing entities are a subject of the UNFSA, in other words, their legal status under the 1995 agreement.

2. Personality in International Law

The subjects of international law are entities with international legal personality which possess the capacity to bear rights and obligations under the international legal system.[1] In the past, states were regarded as the only subject of international law. However, the concept of subject in international law is not invariable. An advisory opinion of the International Court of Justice (ICJ) in *Reparation for Injuries Suffered in the Service of the United Nations* declares that "[t]he subjects of law in any legal system

---

[1] Bin Cheng, "Introduction to Subjects of International Law," in Mohammed Bedjaoui, ed., *International Law: Achievements and Prospects* (Dordrecht: Martinus Nijhoff Publishers, 1991), p. 23.

are not necessarily identical in their nature or in the extent of their rights, and their nature depends upon the needs of the community."[2] The development of international society has created various subjects of international law. This ICJ opinion states that the international organisation "is a subject of international law and capable of possessing international rights and duties, and that it has capacity to maintain its rights by bringing international claims."[3]

As the subjects of international law have varied over time, the capacity which entities possess to bear rights and duties differs for each subject. However, the variations of capacity are not the criteria by which to judge whether entities possess international legal personality. O'Connell explains this point clearly.

> Capacity implies personality, but always it is capacity to do those particular acts. Therefore "personality" as a term is only shorthand for the proposition that an entity is endowed by international law with legal capacity. But entity A may have capacity to perform acts X and Y, but not act Z, entity B to perform acts Y and Z but not act X, and entity C to perform all three. "Personality" is not, therefore, a synonym for capacity to perform acts X, Y and Z; it is an index, not of capacity per se, but of specific and different capacities. … So a State may have capacity to do acts X, Y and Z, the United Nations to do acts Y and Z, the International Labour Organisation to do act X, and the human being to do acts X and Y. All four entities have capacities. To deny that the last three have personality is to argue that only entities with all capacities are persons, an argument that removes all meaning from the term "personality." This was the error made by generations of international lawyers who asserted that "States only are the subjects of international law."[4]

As an international personality cannot be judged by its various capacities, how then can it be decided if an entity possesses international personality? In slave-holding societies, slaves were regarded as human beings but were not endowed by the legal system with rights and duties; in other words, they were the object, instead of the subject, of the law.[5] In contrast, the Roman Emperor Caligula conferred on his horse a Consul

---

[2] *Reparations for Injuries Suffered in the Service of the United Nations, Advisory Opinion: ICJ Reports 1949, p.* 174, 179.

[3] *Ibid.*, p. 178. Also *see* Ian Brownlie, *Principles of Public International Law*, 6th ed. (New York: Oxford University Press, 2003), p. 57.

[4] D.P. O'Connell, *International Law*, 2nd ed. (London: Stevens and Sons, 1970, Vol. 1), pp. 81-82.

[5] Cheng, *supra note* 1, p. 24.

status, with all the attendant legal rights and duties of the Roman office superior to normal citizenship.[6] These examples illustrate that legal personality can be endowed to different types of entities and that the rights and duties that each entity possesses can be different as well.[7] Therefore, an entity possessing legal personality in legal systems A, B and C does not necessarily possess legal personality in legal system X.[8]

An international law system can be created by any two autonomous entities not restrained by any superior legal authority. These entities decide to enter legal relations by making a treaty or contract on the basis of equality and reciprocity, thus establishing an international legal system.[9] These entities possess international legal personality in the international legal system which they have created. Any third party or entity can also possess, or be endowed with, international legal personality in that system provided that the original founding entities agree. As Dr Bin Cheng states:

> Once an international legal system has been established among a limited number of entities, it is obvious that the entrance of new members will require the consent of existing members who, as the legislative and administering authorities of the legal system, will have to be satisfied with the factual ability of the applicant to bear legal rights and duties under the system before granting it the status of a subject of the system.[10]

Consequently, the existing members' approval for a new member to participate in their international legal system indicates that they recognise that the applicant possesses the capability to bear the rights and duties of that system, and this recognition endows the new member with international legal personality in their legal system.

Accordingly, fishing entities are regarded as possessing international legal personality in the international legal system if the existing contracting parties or members view them as having the capability to bear rights and obligations of the legal system and agree that they may become the member. As a result, the decision whether a fishing entity can possess international legal personality crucially depends on recognition from the members of a specific legal system. Traditionally, recognition theories in

---

[6] *Ibid.*

[7] *Ibid.*, p. 25.

[8] Even in a single system, there might be different outcomes: *See* also the infamous *Dred Scott* v. *Sandford* decision by the US Supreme Court. As a slave, Scott was a subject of the law in some states but an object in others.

[9] Cheng, *supra note* 1, p. 31.

[10] *Ibid.*, p. 35.

international law are applied to states or governments. According to constitutive theory, a state or government can become an international person and a subject of international law only after it is recognised by other states; thus, recognition becomes a necessary condition for construction of a state or government.[11] The other doctrine concerning recognition is declaratory theory, according to which the existence of a state or government is a fact, and recognition is an acknowledgement of that fact; hence, a newly established state has a right to be treated as a state regardless of whether it has been recognised by other states.[12]

Fishing entities are entities that perform the function of fishing, that is, they have the capability to fish. However, possessing the capability to fish does not mean they also have the capability to bear the rights and obligations of the international legal system because even a private fishing company or an individual fisherman has that capability. Therefore, recognition becomes a key element in whether fishing entities have international personality. In the case of states or governments, most scholars adopt declaratory theory[13] because constitutive theory could result in the confusing and complicated legal consequence that a state is recognised by State A but not State B and thus simultaneously is and is not an international personality.[14]

However, in the case of fishing entities, different thinking is adopted, and constitutive theory is often applied to fishing entities. When a state exists as a state, it means that it is an entire international person who can bear rights and obligations under international law in various fields, not merely a particular field. Hence, the problem that a state faces is not whether it has international personality in certain fields of particular international legal systems but whether the existence of the state itself is an established fact. However, a fishing entity is different from a state and focuses solely on the function of fishing. Whether it has international personality in the field of international fisheries law cannot depend only on whether the existence of the fishing entity is a fact but also whether it can bear rights and obligations under international fisheries law. Therefore, this capability is a crucial element for a fishing entity to be regarded as possessing international personality. Consequently, recognition by other states in separate international legal systems, *i.e.* international agreements

---

[11] *See* D.J. Harris, *Cases and Materials on International Law*, 7th ed. (London: Sweet & Maxwell, 2010), p. 130.
[12] *Ibid.*
[13] *Ibid.*, p. 131.
[14] *Ibid.*, p. 130.

or instruments, becomes a necessary condition for a fishing entity to be the international person in those legal systems.

In addition, fishing entities face a different situation as the subjects of general international law than states. Although the term "fishing entity" gained attention through the UNFSA, not every international fisheries agreement or instrument confers subject status upon fishing entities. Within those legal systems which do not allow fishing entities to become a subject, fishing entities have no opportunity to possess international legal personality. Thus, it is an overstatement to conclude that a fishing entity can be recognised as possessing international legal personality in the general international fisheries law. In other words, whether a fishing entity has international legal personality can be assessed only in separate legal systems. Only if the legal system confers upon a fishing entity the right to become a subject and if other parties in the legal system recognise that the fishing entity possesses the capability to bear the rights and obligations of the legal system can the fishing entity be regarded as possessing international personality in that legal system.

Regarding Taiwan, a fishing entity in many RFMOs, the Taiwanese government formed after withdrawing from mainland China in 1949 amid a civil war with the Communist Party, which has since controlled mainland China. Both the Beijing and the Taiwan governments claim to be the government of the one state of China. Consequently, they have both sought recognition. The Taiwanese government was recognised as the government of the one state of China until 1971 when its seat in the UN was replaced by the Beijing government. From that time, the Beijing government has been recognised as the government of the one state of China and has claimed Taiwan as part of its territory.[15] The Taiwan government, although not recognised, still claims to be the legal government of China, *i.e.* seeks recognition of itself. In 2000, Taiwanese President, Shui-Bian Chen, described Taiwan as an independent state, and the Taiwan government has sought to join the UN, not as a replacement for the Beijing government, but as the government of a new member state.[16] Although the majority of states have not recognised Taiwan as a state, the

---

[15] For details of the history of the Taiwan and Beijing governments, please *see* Chapter 7 of this book.

[16] *See* UN Doc. A/58/197. Also *see* D.J. Harris, *Cases and Materials on International Law*, 7th ed. (London: Sweet & Maxwell, 2010), p. 96. The Beijing government opposed recognising Taiwan as a state; its full statement can be seen on the website of the Permanent Mission of the People's Republic of China to the UN, available at *http://www.china-un.org/eng/lhghyywj/smhwj/wangnian/fy03/t29409.htm* (visited on 26/03/2014)

Taiwan government has changed its claim from seeking the recognition of its government to recognition of its statehood.

Pursuant to the 1933 Montevideo Convention on Rights and Duties of States[17], the conditions that constitute a state are a permanent population, defined territory, government and the capacity to enter into relations with other states. Taiwan has 23 million inhabitants, the territory of Taiwan and its surrounding islands, the Taiwan government, and formal diplomatic relations with 23 states.[18] Therefore, Taiwan unarguably can be regarded as a full state. In addition, according to declaratory theory, the existence of Taiwan as a state is a fact, so it possesses international personality. However, the majority of other states do not recognise Taiwan as a state due to the Beijing government's claim that Taiwan is part of China; in other words, they have not established formal diplomatic relations with Taiwan, but continue to deal with Taiwan in foreign affairs.[19] Facing opposition from the Beijing government, Taiwan cannot participate in RFMOs under the identity of a state, leading to the present circumstances of its participation in RFMOs under the identity of a fishing entity. However, a fishing entity is different than a state, particularly its legal status. Thus, a fishing entity possesses personality only in certain specific international legal systems.

Whether fishing entities possess international legal personality is usually decided through checking the constitution, charter, treaty or other documents of the legal system and determining whether any provisions or regulations confer on fishing entities the possibility of becoming contracting parties or members.[20] Therefore, each RFMO can be regarded as a separate legal system in which fishing entities might possess different legal status. This trend also applies to each fisheries agreement. Accordingly, we will analyse whether fishing entities possess international legal personality under the following agreements, which are two of the most significant agreements in the international fisheries context.

---

[17] For the full text of the convention, please *see* Malcolm D. Evans, *International Law Documents*, 8th ed. (Oxford: Oxford University Press, 2007), pp. 8–9.

[18] Regarding the 23 states, please *see* Chapter 7, footnote 22, of this book.

[19] The examples are shown in Chapter 7, footnote 23, of this book.

[20] Martin Tsamenyi, "The Legal Substance and Status of Fishing Entities in International Law: A Note," *Ocean Development and International Law*, Vol. 37(2006), p. 125.

3. United Nations Convention on the Law of the Sea

The UNCLOS does not use the term "fishing entities;" however, Article 1, paragraph 2(2) states that "[t]his Convention applies *mutatis mutandis* to the entities referred to in article 305, paragraph l(b), (c), (d), (e) and (f), which become Parties to this Convention in accordance with the conditions relevant to each, and to that extent "states parties" refers to those entities." Article 305 describes the subjects who may sign the convention. In addition to all states and some specific international organisations, the convention is open to other entities, which Article 305, paragraph l(b), (c), (d), (e) and (f) identifies these entities as "Namibia, represented by the United Nations Council for Namibia;" "all self-governing associated States which have chosen that status in an act of self-determination supervised and approved by the United Nations in accordance with General Assembly resolution 1514 (XV) and which have competence over the matters governed by this Convention, including the competence to enter into treaties in respect of those matters;" "all self-governing associated States which, in accordance with their respective instruments of association, have competence over the matters governed by this Convention, including the competence to enter into treaties in respect of those matters;" and "all territories which enjoy full internal self-government, recognized as such by the United Nations, but have not attained full independence in accordance with General Assembly resolution 1514 (XV) and which have competence over the matters governed by this Convention, including the competence to enter into treaties in respect of those matters." These provisions concerning the conditions of ratification, formal confirmation and accession are similar to Article 305. Article 306 states that "[t]his Convention is subject to ratification by States and the other entities referred to in article 305, paragraph l(b), (c), (d) and (e), and to formal confirmation, in accordance with Annex IX, by the entities referred to in article 305, paragraph l(f)," and Article 307 that "[t]his Convention shall remain open for accession by States and the other entities referred to in article 305. Accession by the entities referred to in article 305, paragraph l(f), shall be in accordance with Annex IX."

Although this international instrument does not define the term "fishing entity," it can be regarded as an entity that performs the function of fishing and is recognised as possessing the capability to bear the rights and obligations in certain legal system. Thus, Article 305, paragraph 1(d)(e) of UNCLOS can be applied to fishing entities; in other words, any entity with the function of fishing may, pursuant to Article 305, paragraph 1(d) or (e), apply to become a party to UNCLOS. Such participation confirms that they recognise it as possessing the capability to bear the rights and obligations of the UNCLOS and agree that it possesses international personality as a result of becoming a party under these two articles. In practice, Taiwan may be a fishing entity but not a party to UNCLOS because Taiwan is not a "self-governing associated state." Taiwan might "enjoy full self-government, recognized as such by the UN, but has not attain full independence in accordance with General Assembly resolution 1514(XV)." Although UNCLOS does not clearly state criteria for a fishing entity, any entity that performs the function of fishing and corresponds to Article 305, paragraph 1(d) or (e) of UNCLOS has the possibility of becoming a party to UNCLOS. Nevertheless, no fishing entity may apply to be a party pursuant to these articles.

Obviously, UNCLOS is open for some specific but not all fishing entities may sign or accede to UNCLOS. Fishing entities possibly lack the opportunity to be a subject of UNCLOS, which excludes some fishing entities such as Taiwan from joining.

4. United Nations Fish Stocks Agreement

The UNFSA is the first international agreement with provisions concerning fishing entities. It not only reveals the significance of fishing entities in high seas fishery conservation and management measures but also increases the possibilities for fishing entities to participate in RFMOs. Articles 1(3) and 17(3) introduced fishing entities into this particular legal system. These concepts and provisions gradually developed through successive drafts in the six sessions of the UN Conference on Straddling Fish Stocks and Highly Migratory Fish Stocks.

4.1 The Organisational Session

The UN convened a conference to effectively implement the provisions of Articles 63[21] and 64[22] in UNCLOS concerning straddling fish stocks and

[21] Article 63 of UNCLOS reads:
"1. Where the same stock or stocks of associated species occur within the exclusive economic zones of two or more coastal States, these States shall seek, either directly or through appropriate subregional or regional organizations, to agree upon the measures necessary to coordinate and ensure the conservation and development of such stocks without prejudice to the other provisions of this Part.
2. Where the same stock or stocks of associated species occur both within the exclusive economic zone and in an area beyond and adjacent to the zone, the

highly migratory fish stocks, gradually leading to the UNFSA. During the framing of the agreement, the FAO played an important role. Pursuant to UN General Assembly Resolution 47/192 [23], one task of the UN Conference on Straddling Fish Stocks and Highly Migratory Fish Stocks was to make recommendations based on scientific and technical studies conducted by the FAO.[24]

The Organizational Session of the UN Conference on Straddling Fish Stocks and Highly Migratory Fish Stocks was held in New York 19–23 April 1993.[25] During this session, the FAO representative informed the

---

coastal State and the States fishing for such stocks in the adjacent area shall seek, either directly or through appropriate subregional or regional organizations, to agree upon the measures necessary for the conservation of these stocks in the adjacent area."

[22] Article 64 of UNCLOS reads: "The coastal State and other States whose nationals fish in the region for the highly migratory species listed in Annex I shall cooperate directly or through appropriate international organizations with a view to ensuring conservation and promoting the objective of optimum utilization of such species throughout the region, both within and beyond the exclusive economic zone. In regions for which no appropriate international organization exists, the coastal State and other States whose nationals harvest these species in the region shall cooperate to establish such an organization and participate in its work."

[23] *A/RES/47/192*. The full text can be found at *http://www.un.org/documents/ga/res/47/a47r192.htm* (visited on 28/02/2011).

[24] Paragraph 2 of the resolution states that the General Assembly "decides that the intergovernmental conference, in accordance with the said mandate, shall take into account relevant activities at the subregional, regional and global levels, with a view to promoting effective implementation of the provisions of the United Nations Convention on the Law of the Sea on straddling fish stocks and highly migratory fish stocks, and that it, drawing, inter alia, on scientific and technical studies by the Food and Agriculture Organization of the United Nations, should: (a) Identify and assess existing problems related to the conservation and management of such fish stocks; (b) Consider means of improving fisheries cooperation among States; (c) Formulate appropriate recommendations." This part of the resolution is also mentioned in *A/CONF. 164/7*, Statement Made by the Chairman of the Conference at the Opening of the Organizational Session, as compiled in Jean-Pierre Lévy and Gunnar G. Schram, eds., *United Nations Conference on Straddling Fish Stocks and Highly Migratory Fish Stocks—Selected Documents* (Hague: Kluwer Law International, 1996), p. 40.

[25] Oceans and the Law of the Sea, the UN website of Division for Ocean Affairs and the Law of the Sea, *http://www.un.org/Depts/los/fish_stocks_conference/fish_stocks_conference.htm* (visited on 28/02/2011).

conference that FAO would provide it with several documents[26], including reports and documents of the 1992 Technical Consultation on High Seas Fishing. Of these, the most relative to fishing entities is "International Fishery Bodies: Considerations for High Seas Management[27]." Paragraph 9 of this document states that the effectiveness of fishery bodies might be constrained if "unregulated fishing by non-contracting parties undermines efforts to promote rational resource use by fishery bodies. A country or an entity may opt not to participate in such bodies or may be prevented from doing so for political or other reasons. The impact of fishing by non-contracting parties on management efforts in a fishery can be severe, and in the extreme, is capable of negating attempts to secure sustainable resource use."[28] Here, the term "entity" refers to the entities described in Article 305, paragraphs 1(c), (d) and (e) of UNCLOS as possess the right to become contracting parties[29] In addition, the FAO "International Fishery Bodies Document" may also refer to Taiwan, as proved in paragraph 46 of the document. Paragraph 9 of the document clearly points out that fishing by non-contracting parties, whether a state or another entity, would undermine the conservation and management measures of the RFMOs, implying that this problem should be solved. This consideration is mentioned again in paragraphs 45–47 with references to "non-contracting parties." Paragraph 45 of the document further emphasised the importance of fishing entities' participation in RFMOs; it reads: "Some nations or other entities operating in a fishery may opt not to participate in a high seas management body or they may be excluded from it (e.g., for political or other reasons). The effectiveness of high seas management will therefore be significantly reduced if a major entity in a fishery does not participate in determining management decisions and in turn is not bound by those decisions."[30] Paragraph 46 cites Taiwan as an example of a major

[26] Those documents were reports and documents from the 1992 Technical Consultation on High Seas Fishing; Declaration of Cancun, adopted by the 1992 International Conference on Responsible Fishing, Cancun, Mexico; Strategy for Fisheries Management and Development, adopted by the 1984 World Conference for Fisheries Management and Development; Draft Code of Conduct for Responsible Fishing; and Draft Agreement on Flagging and Reflagging of Fishing Vessels on the High Seas.

[27] The full text of this document is compiled in Jean-Pierre Lévy and Gunnar G. Schram, eds., *supra note* 24, pp. 346-358.

[28] *See* Lévy and Schram, eds., *supra note* 24, p. 348.

[29] Namibia has been an independent state since 1990, so Article 305, paragraph 1(b) of UNCLOS was not included here.

[30] Lévy and Schram, eds., *supra note* 24, p. 355.

fishing entity and argues that "[t]he exclusion of parties from management bodies for political or other reasons poses particular difficulties. Taiwan is a major international fishing entity. Its high seas capacity is extensive and likely to increase, especially in the Indian and South Pacific Oceans. However, due to political non-recognition, Taiwan (Province of China) does not participate fully in any fishery management bodies."[31] Paragraph 47 emphasised that the problems caused by non-contracting parties must be addressed; otherwise, attempts to achieve sustainable use might be thwarted.[32] These principles and concerns indirectly encouraged including fishing entities in the legal framework of international fishery management, particularly through stipulations concerning fishing entities in the UNFSA.

In addition, another important principle of this document suggests that, for effective management of the high seas, RFMOs should possess a significant degree of independence in the execution of their functions.[33] This principle helps each RFMO maintain the independent exercise of authority, whether in the establishment of a commission or decision-making during meetings of the commission.

At the organisation session, a concrete context concerning straddling and highly migratory fish stocks was not discussed. Instead, the rules of procedure and the agenda for the Conference were adopted. Those documents and technical reports prepared by FAO were intended to serve as references in the next session.

4.2 The Second Session

In the second session of the UN Conference on Straddling Fish Stocks and Highly Migratory Fish Stocks held in New York 12–23 July 1993, the

---

[31] *Ibid.*

[32] Paragraph 47 states that "[t]he non-contracting parties problem must be addressed. This is because, despite efforts to manage high seas fisheries, attempts to achieve sustainable use may be thwarted by unregulated fishing by non-contracting parties. Such unregulated activity will erode benefits accruing from measures designed to promote rational exploitation."

[33] Lévy and Schram, eds., *supra note* 24, p. 350. Paragraph 17 of the document reads: "To be effective vehicles for high seas management, fishery bodies need to have a significant degree of independence in the execution of their functions, be assigned powers consistent with their management tasks and receive support from contracting parties. While the following list of issues is not exhaustive in terms of matters that should be addressed to encourage efficiency in management, it covers issues that would, as a minimum, need to be considered."

chairman of the conference prepared and proposed the negotiating text[34] with the purpose of providing a basic text issues discussed by delegates.[35] Paragraph 4 of the Negotiating Text states that "[c]oastal States and States fishing on the high seas shall cooperate to ensure long-term sustainability of straddling fish stocks and highly migratory fish stocks on the high seas. States shall give effect to the duty to cooperate by establishing conservation and management measures for straddling fish stocks and highly migratory fish stocks and commit themselves to responsible fishing, in a manner consistent with the relevant provisions of the United Nations Convention on the Law of the Sea." This paragraph is accompanied by a note stating that any provisions referring to states should be regarded as covering states, the European Economic Community and fishing entities whose vessels fish on the high seas.

Unlike in Article 17(3) of the UNFSA, fishing entities are not mentioned in the provisions concerning "Non-parties to Subregional or Regional Organizations or Arrangements" in paragraphs 35–38 of the Negotiating Text. However, the chairman stressed the importance of the non-parties issue and argued that, as non-parties can undermine conservation and management efforts by international agreements and RFMOs, their roles were of concern to the Conference.[36] He indicated that, in order to achieve sustainable straddling and highly migratory stocks through conservation and management, ways and methods of dealing with new entrants and non-parties should be established.[37]

4.3 The Third Session

The third session of the UN Conference on Straddling Fish Stocks and Highly Migratory Fish Stocks was held in New York 14–31 March 1994.[38] At this session, the Negotiating Text was discussed and modified into the

[34] *A/CONF. 164/13*, compiled Lévy and Schram, eds., *supra note* 24, pp. 73-94. The full text also can be found at *http://daccess-dds-ny.un.org/doc/UNDOC/GEN/N93/702/41/IMG/N9370241.pdf?OpenElement* (visited on 15/02/2011).

[35] Oceans and the Law of the Sea, UN website for the Division for Ocean Affairs and the Law of the Sea, *supra note* 25 (visited on 15/02/2011).

[36] Statement Made by the Chairman of the Conference at the Closing of the Second Session, *A/CONF. 164/15*, compiled in Lévy and Schram, eds., *supra note* 24, pp. 95-99. Also available at *http://daccess-dds-ny.un.org/doc/UNDOC/GEN/N93/443/82/PDF/N9344382.pdf?OpenElement* (visited on 15/02/2011).

[37] *Ibid.*

[38] Oceans and the Law of the Sea, UN website of Division for Ocean Affairs and the Law of the Sea, *supra note* 25 (visited on 17/02/2011).

Revised Negotiating Text[39]. Following Paragraph 4 of Negotiating Text, Paragraph 1 of the Revised Negotiating Text stipulates that "[s]tates have a duty to ensure the long-term sustainability of straddling fish stocks and highly migratory fish stocks." An attached note reads, "For the purposes of these provisions, references to States should be interpreted as including the European Economic Community in matters within its competence. These provisions also apply to the fishing entities whose vessels fish on the high seas."

Fishing entities were not mentioned in stipulations in paragraphs 40–43 on "non-participants in subregional or regional organizations or arrangement." [40] As well, neither the Negotiating nor the Revised Negotiating Text contained provisions regulating signature, ratification, acceptance, approval and accession in a general international convention or agreement. However, the basic structure and content of the 1995 agreement appeared in this Revised Negotiating Text.

4.4 The Fourth Session

At the fourth session held in New York 15–26 August 1994,[41] the Revised Negotiating Text was revised, restructured and reformatted into a document entitled Draft Agreement for the Implementation of the Provisions of the UN Convention on the Law of the Sea of 10 December 1982 Relating to the Conservation and Management of Straddling Fish Stocks and Highly Migratory Fish Stocks[42], which aimed to ensure the long-term conservation and sustainable use of straddling and highly migratory fish stocks.[43] The Draft Agreement was an important milestone in the UN Conference on Straddling Fish Stocks and Highly Migratory

[39] *A/CONF. 164/13/Rev.1*, as compiled in Jean-Pierre Lévy and Gunnar G. Schram, eds., *supra note* 24, pp. 437-461.

[40] *Ibid.*, p. 450.

[41] Oceans and the Law of the Sea, UN website of Division for Ocean Affairs and the Law of the Sea, *supra note* 25 (visited on 17/02/2011).

[42] *A/CONF. 164/22*, compiled in Jean-Pierre Lévy and Gunnar G. Schram, eds., *supra note* 24, pp. 621-652. The full text also can be found at *http://daccess-dds-ny.un.org/doc/UNDOC/GEN/N94/338/70/IMG/N9433870.pdf?OpenElement* (visited on 17/02/2011).

[43] *See* Statement Made by the Chairman of the Conference at the Closing of the Fourth Session, on 26 August 1994, *A/CONF. 164/24*, compiled in Jean-Pierre Lévy and Gunnar G. Schram, eds., *supra note* 24, pp. 653-656. Also available at *http://daccess-dds-ny.un.org/doc/UNDOC/GEN/N94/354/13/PDF/N9435413.pdf?OpenElement* (visited on 17/02/2011).

Fish Stocks because it combined the form and substance of a straddling and highly migratory stocks conservation and management agreement, leading to a model for the 1995 final agreement.

Unlike the Negotiating Text or the Revised Negotiating Text, the Draft Agreement did not include a note describing "states" as the European Economic Community and fishing entities. Rather, the Draft Agreement defines "states parties" as states which have agreed to be bound by it.[44] Furthermore, the Draft Agreement contains provisions concerning fishing entities. Article 1(2) reads, "This Agreement applies *mutatis mutandis* to the entities referred to in article 305, paragraph l(c), (d), (e) and (f), of the Convention which become Parties to this Agreement in accordance with the conditions relevant to each and, to that extent, 'States Parties' refers to those entities."[45] Article 1(3) states that "[t]he relevant principles of the Convention and this Agreement are applicable mutatis mutandis to other fishing entities whose vessels fish on the high seas."[46] This marks the first time that the conference put the term "*mutatis mutandis*" into articles. It continued to employ this usage in a later revised agreement and a final version.

According to Black's Law Dictionary, "*mutatis mutandis*" means "with the necessary changes in points of detail, meaning that matters or things are generally the same, but to be altered when necessary, as to names, offices, and the like."[47] Therefore, this term refers to necessary changes in details, such as the subject, object, name of an institution or title which do not change the original meaning of regulations. In this case, Article 1(3) can be explained as follows: Although the agreement is designed to be applicable to contracting parties, it can also apply to fishing entities whose vessels fish on the high seas.

Fishing entities were not written into the provisions concerning non-participants in subregional or regional fisheries management organisations or arrangements, as well as the Negotiating Text and the Revised Negotiating Text. As a mature agreement compared to the Negotiating Text and the Revised Negotiating Text, the Draft Article includes provisions concerning signature, ratification, acceptance, approval, formal confirmation and accession were contained in Articles 37–39, entering into force in Article 40 and reservations and exceptions in Article 41.

---

[44] *See* Article 1(1)(b) of the Draft Agreement, *A/CONF. 164/22*, *supra note* 42.
[45] *Ibid.*
[46] *Ibid.*
[47] Henry Campbell Black, *Black's Law Dictionary*, 5th ed. (St. Paul Minn. West Publishing Co., 1979) p. 919.

4.5 The Fifth Session

The Draft Agreement was discussed and revised at the fifth session held in New York 27 March–12 April 1995.[48] The provisions define "states parties," and Article 1(2), concerning the entities referred to in Article 305, Paragraph l(c), (d), (e) and (f) of UNCLOS, was reproduced without change in the Revised Draft Agreement[49]. However, Article 1(3) was revised to state that "[t]he provisions of this Agreement shall apply *mutatis mutandis* to other fishing entities whose vessels fish on the high seas." In the Draft Agreement, "the relevant principles of UNCLOS" can apply to fishing entities whose vessels fish on the high seas; however, this stipulation was removed from the Revised Draft Agreement. In the Draft Agreement, "[t]he relevant principles of UNCLOS" should, at least, contain Articles 63–64 of UNCLOS concerning straddling and highly migratory fish stocks. Pursuant to these two articles, distant fishing states and coastal states have the common obligation to cooperate directly or indirectly through RFMOs. Specifically, "indirectly cooperate through RFMOs" could be accomplished through two methods: one, through existing RFMOs and two, through a new RFMO established by both categories of state. Whatever the situation, distant fishing states may not exclude coastal states from participating in the cooperation process; neither may coastal states refuse or prevent distant fishing states participating. Both possess the right and obligation to cooperate simultaneously. In the Revised Draft Agreement, fishing entities whose vessels fish on the high seas cannot access these principles of UNCLOS. The sources of rights that fishing entities could claim clearly were restricted compared to the Draft Agreement, though the limitation might not have made any difference in practice.

In addition to UNCLOS, another change from the Draft Agreement to the Revised Draft Agreement is that "the principles of this Agreement" apply to fishing entities whose vessels fish on the high seas under the former; however, under the latter one, "the provisions of this Agreement" can apply only to fishing entities. This language creates a slight difference.

---

[48] Oceans and the Law of the Sea, UN website of Division for Ocean Affairs and the Law of the Sea, *supra note* 25 (visited on 21/02/2011).

[49] *A/CONF. 164/22/Rev.1*, compiled in Lévy and Schram, eds., *supra note* 24, pp. 671-704. The full text also can be found at *http://daccess-dds-ny.un.org/doc/UNDOC/GEN/N95/105/06/IMG/N9510506.pdf?OpenElement* (visited on 21/02/2011). This revised version has the same title as the Draft Agreement of the Fourth Session. To distinguish between these two documents, the latter is called "Revised Draft Agreement" in this chapter.

"The principles of the Agreement" indicate not only the concrete provisions of the agreement but also its important spirit. Thus, while applying the agreement, the context, comprising any implied and explicit meanings, is applied, in addition to the written provisions. Although this change makes weak texts stronger, it is not necessarily advantageous to fishing entities. Fishing entities usually do not have the same as states within the legal system of fisheries agreements and consequently seek for rights as close to those of states as possible. Therefore, it might be advantageous for fishing entities to draw upon the broader sources of international fisheries law, rather than a fixed regulation, allowing them more flexible in claiming their rights. To a certain extent, then, the change in applying this agreement might further narrow fishing entities' rights.

Furthermore, the provisions of Article 17 in the Revised Draft Agreement entitled "States which are not Members of or Parties to Subregional or Regional Fisheries Management Organizations or Arrangements" are mostly the same as the 1995 agreement. However, the content of Article 17(3) of the 1995 agreement appears for the first time in the Revised Draft Agreement. It reads: "States which are members of, or parties to, a subregional or regional fisheries management organisation or arrangement shall, jointly or individually, request the fishing entities referred to in article 1, paragraph 3 which have fishing vessels in the relevant area, to cooperate fully with such organisation or arrangement in implementing the conservation and management measures of that organisation or arrangement, with a view to having such measures applied *de facto* as extensively as possible to fishing activities in the relevant area. Such fishing entities shall enjoy benefits from participation in the fishery commensurate with their commitment to comply with conservation and management measures in respect of the stock(s)." Article 17(3) implies that fishing entities are non-members or non-parties in RFMOs and should be asked to cooperate with other parties or members. In other words, fishing entities should comply with the conservation and management measures of RFMOs; otherwise, their vessels fishing on the high seas might be hindered by states which are members or parties to RFMOs which take related measures.[50]

---

[50] Article 17(4) states that "[s]tates which are members of, or parties to, a subregional or regional fisheries management organization or arrangement shall exchange information with respect to the activities of fishing vessels which fly the flags of States which are neither members of, nor participate in, the organization or arrangement and which are operating in the fishery for the relevant stock(s). They shall take measures consistent with the Convention and this Agreement to deter

On the other hand, Article 17(3) stipulates that, as fishing entities which cooperate with other parties or members are non-members or non-parties in RFMOs, they cannot possess the "rights" which belonged to a party or member but can receive the "benefits" in the fishery commensurate with their participation. Articles 1(3) and 17(3) do not exclude fishing entities from being parties or members of a RFMO or enjoying the rights of the distant fishing states. However, if they do not become parties or members, they at least should cooperate with RFMOs in order to enjoy the commensurate benefits. This principle concerning fishing entities had matured by this stage and continued to be used in the sixth session.

4.6 The Sixth Session

The sixth session in New York 24 July– 4 August 1995 adopted the UNFSA.[51] Compared to the Revised Draft Agreement, Article 1(3) of the UNFSA changed "[t]he provisions of this Agreement shall apply *mutatis mutandis* to other fishing entities whose vessels fish on the high seas" to "[t]his Agreement applies *mutatis mutandis* to other fishing entities whose vessels fish on the high seas." Article 17(3) of the UNFSA also underwent slight changes that did not affect its meaning.

Although the UNFSA contains provisions concerning fishing entities, the problem of whether fishing entities may become states parties to the UNFSA remains. Pursuant to Article 1(3), all the regulations can also be applied to fishing entities, theoretically including the provisions for signature, ratification and accession. However, Article 37 of the UNFSA states that the agreement is open to signing by all states and "other entities" referred to in Article 1, paragraph 2(b), which means the entities referred to in Article 305, paragraph 1(c), (d), (e) of UNCLOS and the specific international organisations stipulated in Annex IX, Article 1 of UNCLOS. As well, Articles 1(3) and 1, paragraph 2(b) use the term "*mutatis mutandis*" to describe how the UNFSA applies to those entities.[52] Using

---

activities of such vessels which undermine the effectiveness of subregional or regional conservation and management measures."

[51] Oceans and the Law of the Sea, UN website of Division for Ocean Affairs and the Law of the Sea, *supra note* 25 (visited on 03/03/2011).

[52] Article 1, paragraph 2(b) of the 1995 Agreement reads: "This Agreement applies *mutatis mutandis*: (i) to any entity referred to in article 305, paragraph 1 (c), (d) and (e), of the Convention and (ii) subject to article 47, to any entity referred to as an 'international organization' in Annex IX, article 1, of the Convention which

similar provisions distinguishing between "other entities" and fishing entities, Article 37 allows the former but not the later to sign under clear identification. The same condition governs the provisions concerning ratification of and accession to the UNFSA.[53] It is assume that the UNFSA is not open to being signed by fishing entities.[54] Therefore, fishing entities may not become states parties to the UNFSA. However, the UNFSA provisions concerning fishing entities recognised the importance of the concept of fishing entities to RFMOs. The UNFSA helps fishing entities to participate in RFMOs and even to become a subject in the regulations of individual RFMOs.

5. Conclusion

Fishing entities can possess international legal personality in the international legal system if they are viewed and recognised by the existing contracting parties as having the capability to bear the rights and obligations of the legal system and agreed to become members. The UNCLOS allows some specific but not all fishing entities to sign or accede to it. The fishing entity Taiwan cannot match the conditions stipulated by the UNCLOS; thus, it cannot join the UNCLOS as a party. In other words, it cannot be a subject of the UNCLOS. In addition, the UNFSA, although it can apply *mutatis mutandis* to fishing entities, may not be signed by fishing entities. However, the UNFSA provides fishing entities the opportunity to be the subject of agreements which create RFMOs and plays a crucial role in helping fishing entities participate in RFMOs. The next chapter discusses fishing entities' status in several RFMOs.

---

becomes a Party to this Agreement, and to that extent 'States Parties' refers to those entities."

[53] Article 38 of the 1995 Agreement states: "This Agreement is subject to ratification by States and the other entities referred to in article 1, paragraph 2(b)." Article 39 reads: "This Agreement shall remain open for accession by States and the other entities referred to in article 1, paragraph 2(b)."

[54] Hasjim Djalal, "The Emergence of the Concept of Fishing Entities: A Note," *Ocean Development and International Law*, Vol. 37(2006), p. 119.

CHAPTER THREE

THE FISHING ENTITY IN REGIONAL FISHERIES MANAGEMENT ORGANISATIONS

1. Introduction

To support cooperation in the field of regional fisheries resources, several RFMOs were established to conserve and manage those resources. These organisations adopted measures including data collection and exchange, scientific research, catch quotas, conservation and management policies and enforcement. Those measures and procedures require states whose vessels are fishing in the areas covered by the RFMO to cooperate and comply with their regulations. Traditionally, states are the main subjects of international law; thus, RFMOs' regulations are aimed at the states who participate as members. However, the conservation and management of fish stocks usually need each participant to cooperate to reach an effective outcome. Fishing entities which are different from states would be excluded from the conservation and management measures in RFMOs, although they play an important role in such issues. Therefore, each RFMO adopts a different legal gateway to bring fishing entities into its regime.

The purpose of this chapter is to explore fishing entities status in RFMOs by examining several selected, important RFMOs in each ocean area. In chronological order from the date on which the relevant negotiations, these organisations are the WCPFC, ICCAT, IATTC, ISC, CCSBT, NPAFC, IOTC and SEAFO.[1]

---

[1] This chapter does not discuss the Southern Indian Ocean Fisheries Agreement (SIOFA) because it is still an agreement, which was adopted in 2006 and entered into force on 21 June 2012, rather than a RFMO, although its next move might be to establish a new fisheries organisation. The main organ of the SIOFA is the Meeting of Parties which takes place at least once a year and decides on arrangements for carrying out secretariat services or establishing a secretariat. Further information can be found at *http://www.fao.org/fishery/rfb/siofa/en* (visited on 30/04/2013).

2. Western and Central Pacific Fisheries Commission

The Western and Central Pacific Ocean has abundant tuna fisheries resources. Following UNCLOS's entry into force on 16 November 1994, the Pacific Islands Forum Fisheries Agency[2] convened the Multilateral High-Level Conference on the Conservation and Management of Highly Migratory Fish Stocks in the Western and Central Pacific (MHLC) on South Pacific tuna fisheries in the Solomon Islands. It met 1–5 December 1994 with the gaol to promote responsible fishing operations in the South Pacific region.

After the UNFSA was adopted in 1995, MHLC2 was held in the Marshall Islands 10–13 June 1997 and issued the Majuro Declaration. Paragraph 2 of the declaration reads: "[Participants] *Decide* to ensure that the fishing activities in the region are conducted in a manner fully consistent with the respective rights, obligations and responsibilities of coastal States and territories and other States and fishing entities fishing on the high seas in the region under the Convention and the Implementing Agreement."[3] This paragraph not only explicitly mentions "fishing entities" but also puts fishing entities in the same position as states, implying that fishing entities possess equal rights and obligations as coastal states and other states under the UNFSA.

Amid the political conflict between Taiwan and China which both participated in the MHLCs, the drafting of the WCPFC Convention's position on fishing entities did not go smoothly.[4] Through the efforts of the

---

[2] The FFA was established in August 1979 with the aim to help countries sustainably manage their fishery resources within their 200 mile exclusive economic zones. The FFA is an advisory body providing expertise, technical assistance and other support to its members who make sovereign decisions about their tuna resources and participate in regional decision making on tuna management through agencies such as the WCPFC. There are 17 members: Australia, Cook Islands, the Federated States of Micronesia, Fiji, Kiribati, Marshall Islands, Nauru, New Zealand, Niue, Palau, Papua New Guinea, Samoa, Solomon Islands, Tokelau, Tonga, Tuvalu and Vanuatu. Further information can be found on the FFA website, *http://www.ffa.int/about* (visited on 01/11/2010).

[3] Netherlands Institute for the Law of the Sea, *International Organizations and the Law of the Sea: Documentary Yearbook* (Netherlands: Kluwer Law International, 1985, Vol. 1), pp. 171–174.

[4] At that time, Taiwan was a potential fishing entity. Hence, during the MHLCs, it sought to gain equal status for fishing entities as states. At the same time, China tried to prevent any recognition or implication that Taiwan was a state. Therefore, the dispute between these two parties extended from politics to the forum on

chairman of and states in the MHLCs, the Convention on the Conservation and Management of Highly Migratory Fish Stocks in the Western and Central Pacific Ocean (the WCPFC Convention)[5] was adopted on 4 September 2000. The WCPFC Convention deals with fishing entities in Annex I–Article 9(2)[6], instead of in the main body of the convention. Paragraph 1 of Annex I provides that "[a]fter the entry into force of this Convention, any fishing entity whose vessels fish for highly migratory fish stocks in the Convention Area, may, by a written instrument delivered to the depositary, agree to be bound by the regime established by this Convention. Such agreement shall become effective thirty days following the delivery of the instrument."[7] Paragraph 2 reads: "Such fishing entity shall participate in the work of the Commission, including decision-making, and shall comply with the obligations under this Convention. References thereto by the Commission or members of the Commission include, for the purposes of this Convention, such fishing entity as well as Contracting Parties." This provision implies that fishing entities may be members of but not contracting parties to the WCPFC. In fact, during the MHLCs, it was proposed to distinguish the concept of "contracting parties" from that of "member of the commission" in the WCPFC Convention.[8] Under that proposal, a fishing entity which affirmed its acceptance of the convention was categorised as a "member of the commission." Thus, fishing entities could only become members of the commission, not contracting parties. Although this idea was not explicitly stated in the final version of the convention, it can be seen in Paragraph 2 of Annex I.[9]

---

fisheries conservation and management. For more information about the conflict between Taiwan and China during the MHLCs, please *see* Chapter 7 of this book.

[5] The convention is one of the first regional fisheries agreements to be adopted since the conclusion in 1995 of the UN Fish Stock Agreement. The full textof the convention can be found on the WCPFC website, *http://www.wcpfc.int/doc/convention-conservation-and-management-highly-migratory-fish-stocks-western-and-central-pacific-* (visited on 01/11/2010).

[6] Article 9(2) reads: "A fishing entity referred to in the Agreement, which has agreed to be bound by the regime established by this Convention in accordance with the provisions of Annex I, may participate in the work, including decision-making, of the Commission in accordance with the provisions of this article and Annex I."

[7] The full text of the annex, *see supra note* 5.

[8] *See* MHLC/WP.1/Rev.4, available at *http://www.spc.int/coastfish/Asides/conventions/MHLC/mhlc_sep99.htm* (visited on 01/11/2010). Also *see* Chapter 7 of this book.

[9] In the West and Central Pacific Fisheries Commission Boarding and Inspection Procedure (WCPFC Boarding and Inspection Procedure) adopted in 2006, it can

Paragraph 3 of Annex I explains the method to solve disputes over the interpretation of the convention involving fishing entities: "If a dispute concerning the interpretation or application of this Convention involving a fishing entity cannot be settled by agreement between the parties to the dispute, the dispute shall, at the request of either party to the dispute, be submitted to final and binding arbitration in accordance with the relevant rules of the Permanent Court of Arbitration." The final paragraph of the Annex implies that its provisions do not set a precedent for other RFMOs: "The provisions of this Annex relating to participation by fishing entities are solely for the purposes of this Convention."

During MHLC7, the "Arrangement for the Participation of Fishing Entities" [10] was drafted through informal consultations between the chairman and the delegation of Taiwan. The purpose of the arrangement was to invite and reassure Taiwan that, as a fishing entity, it could "participate in the Preparatory Conference established by the resolution attached to the Final Act of the Conference" and, "subject to the fulfilment of its domestic legal requirements, agree to be bound by the regime established by the Convention in accordance with article 9, paragraph 2, of the Convention, and participate in the Commission for the Conservation and Management of Highly Migratory Fish Stocks in the Western and Central Pacific Ocean in accordance with the Convention." Taiwan signed this arrangement as a fishing entity under the name of "Chinese Taipei" on 5 September 2000.[11]

After another four annual preparatory conferences, the convention established the WCPFC in 2004 with the aim of ensuring, through effective management, the long-term conservation and sustainable use of highly migratory fish stocks in the Western and Central Pacific Ocean in accordance with UNCLOS and the UNFSA.

---

clearly be seen that fishing entities are categorised as members of the commission, differently than contracting parties. The full text of the procedure is available at *http://www.wcpfc.int/doc/cmm-2006-08/western-and-central-pacific-fisheries-commission-boarding-and-inspection-procedures* (visited on 05/01/2012). Regarding fishing entities in the procedure, please *see* Chapter 5, Chapter 6 and Chapter 7 of this book.

[10] The full text of the arrangement can be found on the WCPFC website, *http://www.wcpfc.int/doc/arrangement-participation-fishing-entities* (visited on 01/11/2010).

[11] Michael W. Lodge, "The Practice of Fishing Entities in Regional Fisheries Management Organizations: The Case of the Commission for the Conservation and Management of Highly Migratory Fish Stocks in the Western and Central Pacific Ocean," *Ocean Development and International Law*, Vol. 37(2006), p. 198.

3. International Convention for the Conservation of Atlantic Tunas

Following the adoption of the International Convention for the Conservation of Atlantic Tunas (the ICCAT Convention)[12] in Rio de Janeiro, Brazil, in 1966, the ICCAT, which is responsible for the conservation of tunas and tuna-like species in the Atlantic Ocean and adjacent seas, was established in 1969.[13] The ICCAT Convention does not have any regulations concerning fishing entities; it only allows the UN members and specialized agencies to send observers to participate meetings of the commission.[14] There was no gateway for a fishing entity to participate formally in the ICCAT until 1997.

At the 9th Special Meeting of the ICCAT in 1994, the Resolution on Coordination with Non-Contracting Parties was adopted for the purpose of contacting all non-contracting parties known to be fishing in the convention area for species under the competence of the convention in order to urge them to become contracting parties or cooperating parties which could attend ICCAT meetings as observers.[15] However, the 1994 resolution still does not mention any cooperation specifically related to fishing entities. In November 1997, the Resolution by ICCAT on Becoming a Cooperating Party, Entity or Fishing Entity was adopted by the Commission at the 15th Regular Meeting in Madrid, Spain, and entered into force on 13 June 1998.[16] The 1997 resolution expands the category of non-contracting parties from "cooperating parties" in the 1994 resolution to "cooperating party, entity or fishing entity." This was the first ICCAT

---

[12] The full text of the convention can be found on its official website, *http://www.iccat.int/Documents/Commission/BasicTexts.pdf* (visited on 09/02/2011).

[13] The ICCAT directly concerns approximately 30 species: Atlantic bluefin, skipjack, yellowfin, albacore and bigeye tuna; swordfish; billfishes such as white marlin, blue marlin, sailfish and spearfish; mackerels such as spotted Spanish mackerel and king mackerel; and small tunas such as black skipjack, frigate tuna, and Atlantic bonito. Further information can be found on the ICCAT website, *http://www.iccat.int/en/contracting.htm* (visited on 21/12/2010).

[14] *See* Article XI(3): "The commission may invite any appropriate international organization and any Government which is a member of the United Nations or of any Specialized Agency of the United Nations and which is not a member of the Commission, to send observers to meetings of the Commission and its subsidiary bodies."

[15] *See* the Report of the 9th Special Meeting of ICCAT, available at *http://www.iccat.int/Documents/BienRep/REP_EN_94-95_I_1.pdf* (visited on 21/12/2010).

[16] *See* the Report of the 15th Regular Meeting of ICCAT, available at *http://www.iccat.int/Documents/BienRep/REP_EN_96-97_II_1.pdf* (visited on 21/12/2010).

regulation with provisions explicitly on fishing entities. The purpose of this resolution was to continue to encourage all non-contracting parties, entities and fishing entities with vessels fishing for the ICCAT species to implement the ICCAT conservation measures.[17] Pursuant to paragraph 2 of the 1997 resolution, any non-contracting parties, entities and fishing entities that would like to acquire the status of cooperating party, entity or fishing entity in the ICCAT may apply annually to the executive secretary.

Chinese Taipei, as an observer at the ICCAT, proposed to amend the 1997 resolution at the 12th Special Meeting of the ICCAT in 2000.[18] The proposal by Chinese Taipei suggested that, while recommending to the commission whether an applicant could be considered a cooperating party, entity or fishing entity, the Permanent Working Group for the Improvement of ICCAT Statistics and Conservation Measures (PWG) should consider granting the status of a cooperating party, entity or fishing entity for no more than three years for the applicant entitled to become a member of the commission.[19] Furthermore, paragraph 4 of the proposed amendment would automatically renew cooperating party, entity or fishing entity status annually, instead of requiring separate applications every year, unless revoked by the commission due to non-compliance with the ICCAT conservation and management measures or the commission received a written request for withdrawal of the status.[20] This proposal would be beneficial to the ICCAT objective to obtain the cooperation from non-contracting parties because automatically renewing cooperating party, entity or fishing entity status could further stabilise the relationship between the ICCAT and those non-contracting parties, extending their cooperation period rather than limiting it to one year. However, this proposal was not adopted but slated to be reconsidered at the next meeting.

At the 17th Regular Meeting of the ICCAT in November 2001, the People's Republic of China proposed another draft, the revised 1997

---

[17] *Ibid.*

[18] The full text of the proposal can be found in the 12th Special Meeting of ICCAT, *http://www.iccat.int/Documents/BienRep/REP_EN_00-01_I_1.pdf* (visited on 29/12/2010). Chinese Taipei began to attend meetings of the ICCAT as an observer in 1972. At that time, the concept of the fishing entity had not developed, and Chinese Taipei regarded itself as a state at the ICCAT meetings. The statehood of Chinese Taipei was raised as an issue after the People's Republic of China acceded to the ICCAT Convention in 1996. For a discussion the attendance of Chinese Taipei as an observer in 1972 and the issue of its statehood as related to its conflict with the People's Republic of China, *see* Chapter 8 of this book.

[19] *Ibid.*

[20] *Ibid.*

ICCAT Resolution by on Becoming a Cooperating Party, Entity or Fishing Entity to amend the 1997 resolution.[21] This amendment modifies paragraph 4 of the 2000 proposal by Chinese Taipei, while the rest of the provisions are almost the same as the 1997 resolution.[22] The 2001 amendment was adopted to replace the 1997 resolution and entered into force on 21 September 2002.

In 2003, at the 18th Regular Meeting of the ICCAT, a Recommendation on Criteria for Attaining the Status of Cooperating non-Contracting Party, Entity or Fishing Entity in ICCAT (Recommendation 03-20) was discussed and adopted by consensus to replace the 2001 amendment.[23] Recommendation 03-20 nearly duplicates the stipulations of the 2001 amendment but requires the PWG to "consider information regarding the applicant available from other Regional Fisheries Management Organizations (RFMOs) as well as data submission of the applicant to the Commission. Caution shall be used so as not to introduce into the Convention area the excessive fishing capacity of other regions or IUU fishing activities in granting Cooperating Status to the applicant" while reviewing a request for cooperating status.[24] Furthermore, applicants requesting the status of cooperating non-contracting party, entity or fishing entity are required to confirm their commitment to respect the commission's conservation and management measures and to inform ICCAT of the measures it takes to ensure that its vessels comply with ICCAT conservation and management measures.[25] Under paragraph 3 of Recommendation 03-20, the applicant should also provide the following information:

> "a) where available, data on its historical fisheries in the Convention area, including nominal catches, number/type of vessels, name of fishing vessels, fishing effort and fishing areas;
> b) all the data that Contracting Parties have to submit to ICCAT based on the Recommendations adopted by ICCAT;
> c) details on current fishing presence in the Convention area, number of vessels and vessel characteristics and;

---

[21] *See* the Report of the 17th Regular Meeting of ICCAT, available at *http://www.iccat.int/Documents/BienRep/REP_EN_00-01_II_1.pdf* (visited on 29/12/2010).
[22] *Ibid.*
[23] *See* the Report of the 18th Regular Meeting of ICCAT, available at *http://www.iccat.int/Documents/BienRep/REP_EN_02-03_II_1.pdf* (visited on 29/12/2010).
[24] *See* Paragraph 5 of the Recommendation 03-20.
[25] *Ibid.*

d) information on any research programs it may have conducted in the Convention area and the information and the results of this research."[26]

In addition to the conditions for applicants, the ICCAT lays out the obligations for cooperating non-contracting party, entity or fishing entity in the Recommendation by the ICCAT Concerning the Duties of Contracting Parties and Cooperating non-Contracting Parties, Entities or Fishing Entities in Relation to their Vessels Fishing in the ICCAT Convention Area (Recommendation 03-12). Cooperating non-contracting parties, entities, fishing entity and contracting parties must effectively control their vessels fishing in the ICCAT convention area, including authorising their vessels to fish in the ICCAT convention area only with fishing authorisations, licenses or permits; ensuring that vessels are authorised to fish in the ICCAT convention area only if they can effectively perform their responsibilities, including monitoring and controlling their fishing activities; ensuring that their vessels do not conduct unauthorised fishing in areas under the jurisdiction of other states; requiring their vessels fishing on the high seas to carry their license, authorisation or permit on board at all times and produce it on demand for inspection by a duly authorised person; investigating and following on alleged violations by vessels and reporting the results of such investigations; establishing and maintaining an up-to-date record of fishing vessels; and ensuring that its fishing vessels authorised to fish species managed by ICCAT in the convention area, as well as their fishing gears, are marked in such a way that they can be readily identified in accordance with generally accepted standards, such as the FAO standard specification for the marking and the identification of fishing vessels.[27] Although fishing entities cannot become contracting parties, their vessels conducting fishing in the convention area have no different obligations under Recommendation 03-12.

Although the ICCAT does not distinguish between contracting parties and members in the way the WCPFC does, it offers fishing entities the opportunity to participate in the operation of the ICCAT through the creation of the status of cooperating non-contracting party, entity or fishing entity. However, in the ICCAT, fishing entities can only attend meetings under the identity of observers, which is not a secure status as in the WCPFC. The WCPFC allows fishing entities to become members, which is a status different but equal to that of state members who are contacting parties; furthermore, fishing entities' qualifications to

[26] *Ibid.*
[27] *Ibid.*

participate do not need to be reviewed annually. The ICCAT needs the cooperation of fishing entities or non-contracting parties to effectively carry out its conservation and management measures, but its cooperation with fishing entities is based on a relationship in which the ICCAT sets and reviews the fishing entities' qualifications. However, the effectiveness of ICCAT conservation and management regimes need the compliance of each actor whose vessels conduct fishing in the convertion area. From this perspective, fishing entities appear to be in the dominant, or at least the significant, position; their need to cooperate with the ICCAT is less than the ICCAT's need to cooperate with them. Thus, the ICCAT has placed itself in an unrealistic position facing fishing entities or non-cooperating parties which is not beneficial for the effectiveness of its conservation and management measures.

4. Inter-American Tropical Tuna Commission

The IATTC was set up by the Convention for the Establishment of an Inter-American Tropical Tuna Commission[28], which was signed by the United States and Costa Rica in 1949 for the purposes of conserving and managing tuna and other marine resources in the eastern Pacific Ocean and enhancing scientific research and cooperation concerning these resources. In 1998, it was decided that, since the 1949 IATTC Convention had existed for almost 50 years, it should be changed to take into account the relevant principles of international law related to the conservation and management of living marine resources, such as the UNCLOS, 1992 Agenda 21[29] and Rio Declaration[30], 1993 FAO Agreement to Promote

---

[28] Further information about the IATTC can be found on its official website, *http://www.iattc.org/HomeENG.htm* (visited on 10/11/2010).

[29] Agenda 21 is a UN action plan related to sustainable development which came out of the UN Conference on Environment and Development (UNCED) held in Rio de Janerio, Brazil, in 1992. The plan is a comprehensive blueprint for action to be taken globally, nationally and locally by organisations of the UN, governments and major groups in every area in which humans directly affect the environment. The full text of Agenda 21 is available at the UN website, *http://www.un.org/esa/sustdev/documents/agenda21/english/Agenda21.pdf* (visited on 15/02/2011).

[30] The Rio Declaration is a short document produced at the 1992 UNCED. It lays out 27 principles intended to guide future sustainable development around the world. The full text can be found on the official website of UN Environment Programme, *http://www.unep.org/Documents.Multilingual/Default.asp?DocumentID=78&ArticleID=1163* (visited on 15/02/2011).

Compliance with International Conservation and Management Measures by Fishing Vessels on the High Seas[31], 1995 FAO Code of Conduct for Responsible Fisheries, and UNFSA. Therefore, an *ad hoc* working group to review the 1949 convention was established in 1998.[32]

The 1949 convention did not create an opportunity for fishing entities to participate.[33] In the second session of the working group in January 1999, the Nicaraguan delegate proposed that the term "parties" should cover three categories: states, regional economic integration organisations and separate customs territories. Meanwhile, Venezuela proposed that it cover states and regional economic integration organisations.[34] Nicaragua's proposal was supported by Costa Rica and El Salvador but received some questions as well.[35] The Ecuadorian delegate noted that the term "separate customs territories" was derived from the WTO, which was designed to deal with trade issues. As such, the term might be inappropriate for use in a fishery resources management organisation.[36] The French delegate referred to the term "fishing entity," as provided for in the UNFSA.[37]

As a result of the discussions in the second, third and fourth sessions, Chairman Ambassador Pulvenis in the fifth session in September 2000

---

[31] The full text can be found on the official FAO website, *http://www.fao.org/DOCREP/MEETING/003/X3130m/X3130E00.HTM* (visited on 15/02/2011).

[32] *See* Resolution 98-02, "Resolution on the Establishment of a Working Group to Review the IATTC Convention," June 1998, available on the IATTC website, *http://www.iattc.org/PDFFiles/C-98-02%20Convention%20WG%20resolution%20Jun%2098.pdf* (visited on 10/11/2010).

[33] The only article concerning application for membership to the IATTC is Article V(3): "Any government, whose nationals participate in the fisheries covered by this Convention, desiring to adhere to the present Convention, shall address a communication to that effect to each of the High Contracting Parties. Upon receiving the unanimous consent of the High Contracting Parties to adherence, such government shall deposit with the Government of the United States of America an instrument of adherence which shall stipulate the effective date thereof. The Government of the United States of America shall furnish a certified copy of the Convention to each government desiring to adhere thereto. Each adhering government shall have all the rights and obligations under the Convention as if it had been an original signatory thereof."

[34] *See* Dustin Kung-Hsiung Wang, "Taiwan's Participation in Regional Fisheries Management Organizations and the Conceptual Revolution on Fishing Entity: The Case of the IATTC," *Ocean Development and International Law*, Vol. 37(2006), p. 210.

[35] *Ibid.*

[36] *Ibid., p. 211.*

[37] *Ibid.*

proposed a "Main Pending Issues" paper, which presented four options for the definition of the term "parties:"[38]

> "Option A:
> 1. Delete the brackets[39] and consider that the entities mentioned can be Parties to the Convention;
> 2. Agree [on] the most appropriate term for referring to such entities:
> — '*different* [should be separate] *customs territories'*
> — '*entities', 'fishing entities', 'entities/fishing entities'*.
> Option B:
> 1. Based on the tentative formula in the draft Convention for the Central and Western Pacific (April 2000), introduce a distinction between the Parties, strictly speaking, and the members of the Commission.
> 2. Introduce a new paragraph, to read as follows:
> *'"Members of the Commission" means the Parties to this Convention as well as any [entity] [fishing entity] [entity/fishing entity] that has formally expressed its acceptance of the regime established by this Convention'.*
> 3. Consider the advisability of adding, at the end of this new paragraph, a provision likewise based on the draft Convention for the Central and Western Pacific, as follows:
> *'The status of Member of the Commission in no way prejudice the legal or political status of that [entity] [fishing entity] [entity/fishing entity]'.*
> Option C:
> 1. Introduce in the Convention the concept of *'Non-Parties/entities/ collaborating fishing entities'*, already used in ICCAT, with the insertion of a new paragraph reading as follows:
> *'Non-Parties/entities/collaborating fishing entities' means any State or [entity] [fishing entity][entity/fishing entity] that has formally expressed its acceptance of the regime established by this Convention and its intention to collaborate actively with the Commission'.*
> 2. Proceed to the necessary adjustments to the text of the Convention, in particular the provisions relating to observers.
> Option D:
> Delete the words in brackets in the Revised Chairman's Text, not introduce any specific provision regarding this situation and let it be regulated by the rules of general international law, and by the decisions adopted by the Commission, as appropriate, in the light of the provisions of the 1995 [UNFSA] and of current practice on this issue (in particular in the framework of ICCAT)."

---

[38] *Ibid.*, pp. 212–213.

[39] Before this paper, in 1999, the Mexican delegate suggested bracketing the definition of parties and discussing it after solving other issues. For details, *see* Chapter 7of this book.

Many delegates were inclined to adopt Option B, or the WCPFC model, as reflected in the sixth session of the Working Group in March 2001. There, the chairman presented the Chairman's Consolidated Text which divided the participants into two categories: contracting parties, which included states and regional economic integration organisations, and members of the commission, which would be contracting parties and fishing entities.[40] At the ninth session of the working group in September 2002, a Revised Consolidated Text provided by the chairman confirmed that the WCPFC model would be the final version.[41] It reads:[42]

> "Article I. Definitions
> 6. "Parties" means the States and regional economic integration organizations which have consented to be bound by this Convention and for which this Convention is in force, in accordance with the provisions of Articles XXVII, XXIX, and XXX;
> 7. "Members of the Commission" means the Parties and any fishing entity which has expressed in accordance with the provisions of Article XXVIII its formal commitment to abide by the terms of this Convention and comply with any conservation and management measures adopted pursuant thereto."

The completed draft of the Antigua Convention by the working group was adopted at the 70th meeting of the IATTC on 26–28 June, 2003.[43] At the same meeting, the commission adopted a Resolution on the Participation of a Fishing Entity in the Antigua Convention, which reads:[44]

> "….*Recognizing* its active participation as an observer under the name Taiwan in the work of the Inter-American Tropical Tuna Commission, due to its significant presence in the Eastern Pacific Ocean for over 30 years, and in the negotiation leading to the adoption of the Antigua Convention;
> *Calls upon* that observer to sign the instrument and/or provide the written communication referred to in Article XXVIII of the Antigua Convention, drafted in accordance with the texts attached to the Resolution on the adoption of the Antigua Convention, in its character as a fishing entity under the name Chinese Taipei."

---

[40] Wang, *supra note* 34, p. 213.
[41] *Ibid., p. 215.*
[42] *Ibid., pp. 215-216.*
[43] The Record of Decision, Resolution C-03-02, is available on the IATTC website, *http://www.iattc.org/PDFFiles2/Antigua%20Convention%20-%20Record%20of%20decision.pdf* (visited on 10/11/2010).
[44] Resolution C-03-09, available at *http://www.iattc.org/PDFFiles2/Resolutions/C-03-09%20Participation%20of%20fishing%20entity.pdf* (visited on 10/11/2010).

Articles I(6) and I(7) define "parties" and "members of the Commission," respectively in nearly the same terms as the Revised Consolidated Text.[45] Article XIX stipulates that Article XVIII concerning implementation, compliance and enforcement by parties applies *mutatis mutandis*, to fishing entities that are members of the commission.[46] Similarly, Article XXI provides that Article XX concerning the duties of flag states applies *mutatis mutandis* to fishing entities that are members of the commission. Article XXVI(1) encourages the fishing entities mentioned in article XXVIII to become members or to adopt laws and regulations consistent with this convention. Article XXVIII is an important provision for fishing entities and provides the approach for fishing entities to participate the commission:[47]

> 1. Any fishing entity whose vessels have fished for fish stocks covered by this Convention at any time during the four years preceding the adoption of this Convention may express its firm commitment to abide by the terms of this Convention and comply with any conservation and management measures adopted pursuant thereto, by:
> (a) signing, during the period referred to in Article XXVII, paragraph 1 of this Convention, an instrument drafted to this effect in accordance with a resolution to be adopted by the Commission under the 1949 Convention; and/or
> (b) during or after the above-mentioned period, providing a written communication to the Depositary in accordance with a resolution to be adopted by the Commission under the 1949 Convention. The Depositary shall promptly provide a copy of this communication to all signatories and Parties.
> 2. The commitment expressed pursuant to paragraph 1 of this Article shall be effective from the date referred to in Article XXXI, paragraph 1, of this Convention, or on the date of the written communication referred to in paragraph 1 of this Article, whichever is later.

---

[45] Article I (6) reads: "Parties" means the States and regional economic integration organizations which have consented to be bound by this Convention and for which this Convention is in force, in accordance with the provisions of Articles XXVII, XXIX, and XXX of this Convention; Article I (7) reads: "Members of the Commission" means the Parties and any fishing entity which has expressed in accordance with the provisions of Article XXVIII of this Convention its formal commitment to abide by the terms of this Convention and comply with any conservation and management measures adopted pursuant thereto.

[46] The full text can be found on the IATTC website, *http://www.iattc.org/PDFFiles2/Antigua_Convention_Jun_2003.pdf* (visited on 10/11/2010).

[47] *Ibid.*

> 3. Any fishing entity referred to above may express its firm commitment to abide by the terms of this Convention as it may be amended pursuant to Article XXXIV or Article XXXV of this Convention by providing a written communication to this effect to the Depositary in accordance with the resolution referred to in paragraph 1 of this Article.
> 4. The commitment expressed pursuant to paragraph 3 of this Article shall be effective from the dates referred to in Article XXXIV, paragraph 3, and Article XXXV, paragraph 4, of this Convention, or on the date of the written communication referred to in paragraph 3 of this Article, whichever is later.

Along with the UNFSA, the Antigua Convention uses the term "*mutatis mutandis*" to express that fishing entities are under the same obligations as parties. It is worth noting that, although Articles XVIII and XX are in Part IV on the Rights and Obligations of Members of the Commission, the subject they address is the party, rather than the member. Articles XIX and XXI clearly stipulate that the former two articles apply to fishing entities that are members of the IATTC. This implies that, although fishing entities cannot become contracting parties, they still have status equal to contracting parties.[48] Furthermore, although the Antigua Convention adopts the model of WCPFC distinguishing between parties and members of the commission, the related provisions concerning fishing entities are included in the main text of the Antigua Convention but in the annexes of the WCPFC Convention.

Pursuant to the Antigua Convention, Chinese Taipei, as a fishing entity, signed an Instrument for the Participation of Fishing Entities in Washington, D.C., on 14 November 2003 and became a member of the IATTC when the Antigua Convention entered into force on 27 August 2010.[49]

5. International Scientific Committee for Tuna and Tuna-like Species in the North Pacific Ocean

In 1995, the United States and Japan announced their intention to establish an interim scientific committee (ISC) to study the tuna and tuna-like

---

[48] In the WCPFC Convention, the articles concerning the obligations of Members, duties of flag states and compliance and enforcement all refer to the subject of members, rather than contracting parties. *See* articles 23, 24 and 25 of the WCPFC Convention.

[49] *See* the IATTC website, *http://www.iattc.org/IATTCdocumentationENG.htm* (visited on 17/02/2011).

species of the North Pacific Ocean.[50] The main purpose of the committee would be to increase scientific research and cooperation for conservation and rational utilisation of the species of tuna and tuna-like fishes which inhabit the North Pacific Ocean. Attachment 4 of the Report of the First Meeting of the Interim Scientific Committee for Tuna and Tuna-like Species in the North Pacific contains the Guidelines for the Committee, which was developed through a series of consultations between the United States and Japan.[51] In Part B of the guidelines, the membership is limited to members and observer participants. The former includes coastal states in the region and states with vessels fishing for tuna and tuna-like species in the region, while the latter consists of relevant intergovernmental fishery organisations, intergovernmental marine science organisations and other entities with vessels fishing for tuna and tuna-like species in the region.[52]

According to the original guidelines, only states may be members, and fishing entities may only become observers of the ISC. In 2002, an amendment was proposed which would allow fishing entities to become members. The item concerning other entities with vessels fishing for tuna and tuna-like species in the region in the section addressing observer participants would be deleted, and the phrases "fishing entities" and "fishing entities with vessels fishing for tuna and tuna-like species in the region" added to the section on members (see Table3-1).[53] According to this amendment to the ISC guidelines, fishing entities could be qualified as members rather than observer participants. This proposal was adopted at the 3rd ISC Plenary Meeting on 29 January, 2002, and the revised guidelines become the present version.[54]

As a result, the ISC recognises no difference between contracting parties and members; states and fishing entities which would like to participate in the ISC are categorised as the members. Thus, within the

---

[50] In 2005, the committee name was changed to the "International Scientific Committee for Tuna and Tuna-like Species in the North Pacific Ocean," as recorded in paragraph 12 of the Report of the Plenary Session of the Fifth Meeting of the Interim Scientific Committee for Tuna and Tuna-like Species in the North Pacific. *See* ISC, *http://isc.ac.affrc.go.jp/pdf/ISC5pdf/Report_ISC5_Plenary.pdf* (visited on 07/10/2010).

[51] For further information, *see* ISC, *http://isc.ac.affrc.go.jp/pdf/ISC1pdf/isc1P_rep.pdf* (visited on 07/10/2010).

[52] *Ibid.*

[53] Further information can be found on the ISC website, *http://isc.ac.affrc.go.jp/pdf/ISC3pdf/isc3P_rep.pdf* (visited on 07/10/2010).

[54] *Ibid.*

ISC, fishing entities have equal legal status with states; both state members and fishing entity members have the same rights and obligations. In the light of the revised guidelines, Chinese Taipei, as a fishing entity, announced on 29 January 2002 that, as the soon-to-be-member of the ISC, Taiwan would regard itself as a constructive force in the ISC.[55]

Table 3-1

The Original Provision of Part B	**The 2002 Amendment**
B. Membership 1. Member a. Coastal states of the region b. States with vessels fishing for these species in the region 2. Observer participants a. Relevant intergovernmental fishery organisations b. Relevant intergovernmental marine science organisations c. Other entities with vessels fishing for those species in the region	B. Membership 1. Member a. Coastal states/ **fishing entities** of the region b. States/ **fishing entities** with vessels fishing for these species in the region 2. Observer participants: a. Relevant intergovernmental fishery organisations b. Relevant intergovernmental marine science organisations c. **(Deleted)**

Source: ISC, *http://isc.ac.affrc.go.jp/pdf/ISC3pdf/isc3P_rep.pdf* (visited on 07/10/2010)

[55] The statement can be found in Attachment 3 of Report of the Plenary Session of the Third Meeting of the ISC, *http://isc.ac.affrc.go.jp/pdf/ISC3pdf/isc3P_rep.pdf* (visited on 10/09/2010).

6. Commission for the Conservation of Southern Bluefin Tuna

In May 1993, Australia, Japan and New Zealand signed the Convention for the Conservation of Southern Bluefin Tuna[56], which entered into force on 20 May 1994 and lead to the establishment of the CCSBT headquartered in Canberra, Australia.[57] The southern bluefin tuna (SBT) is the only stock the CCSBT manages and conserves, and the CCSBT's purpose is to ensure, through appropriate management, the conservation and optimum utilisation of this stock.[58]

Article 18 of the convention states that, "[a]fter the entry into force of this Convention, any other State, whose vessels engage in fishing for southern bluefin tuna, or any other coastal State through whose exclusive economic or fishery zone southern bluefin tuna migrates, may accede to it. This Convention shall become effective for any such other State on the date of deposit of that State's instrument of accession."[59] Article 18 stipulates the qualifications for states to accede but does not directly allow the possibility for fishing entities to participate. However, three articles in the convention mention "entity."

Article 5(4) states that "[t]he Parties shall cooperate in the exchange of information regarding any fishing for southern bluefin tuna by nationals, residents and vessels of any State or entity not party to this Convention."[60] Article 14(1) allows the commission to invite states or entities which are not contracting parties but whose nationals, residents or fishing vessels harvest SBT to send observers to meetings of the commission and of the Scientific Committee.[61] Additionally, Article 15(1) reads that "[t]he Parties agree to invite the attention of any State or entity not party to this Convention to any matter relating to the fishing activities of its nationals, residents or vessels which could affect the attainment of the objective of this Convention." Article 15(2) states that "[e]ach Party shall encourage its

---

[56] The full text of the convention can be found on the CCSBT website, *http://www.ccsbt.org/docs/pdf/about_the_commission/convention.pdf* (visited on 15/12/2010).
[57] *See* the website of CCSBT, *http://www.ccsbt.org/docs/about.html* (visited on 15/12/2010).
[58] *See* Article 3 of the Convention, *supra note* 56. Pursuant to this article, South Korea and Indonesia joined the commission on 17 October 2001 and 8 April 2008, respectively.
[59] *Ibid.*
[60] *Ibid.*
[61] *Ibid.*

nationals not to associate with the southern bluefin tuna fishery of any State or entity not party to this Convention, where such association could affect adversely the attainment of the objective of this Convention." Article 15(4) again mentions "entity:" "The Parties shall cooperate in taking appropriate action, consistent with international law and their respective domestic laws, to deter fishing activities for southern bluefin tuna by nationals, residents or vessels of any State or entity not party to this Convention where such activity could affect adversely the attainment of the objective of this Convention."[62]

From these three articles and Article 18, it is obvious that the convention does not consider whether fishing entities could apply to be contracting parties. In addition, the convention does not give criteria for fishing entities to become members of the commission. Instead, pursuant to Article 14(1), the only route for fishing entities to participate the CCSBT is to become observers, provided they are invited by the commission.

Nevertheless, at its seventh meeting in April 2001, the CCSBT adopted a Resolution to Establish an Extended Commission and an Extended Scientific Committee.[63] The objective of the resolution was to ensure, through the commission, all states and fishing entities fishing SBT could work together in sustaining this stock. Paragraph 1 states that part of the legal foundation of the resolution is Article 8(3)(b) of the convention, which provides that the commission can decide and adopt other measures for the purpose of conservation, management and optimum utilisation of SBT. The other legal basis is Article 15(4) of the convention, which encourages all parties to take appropriate action to deter fishing activities for SBT by non-party states and fishing entities which might undermine the objectives of the convention. Hence, instead of the CCSBT proper, the CCSBT's Extended Commission is open to all states and fishing entities

---

[62] *Ibid.*

[63] *See* Attachment I of the Report of the Seventh Annual Meeting of the Commission, available at *http://www.ccsbt.org/docs/pdf/meeting_reports/ccsbt_7/report_of_ccsbt7.pdf* (visited on 15/12/2010). The CCSBT adopted this resolution with the hope of bringing Taiwan, as a major catcher of SBT in the CCSBT Convention area, into the CCSBT regime. Taiwan regarded itself as a state and preferred to participate at least as a member rather than an observer. However, the conflict between Taiwan and China became the main obstacle for Taiwan's participation as a state. Hence, the CCSBT adopted this resolution in order to provide another route for Taiwan to participate in the CCSBT as a fishing entity. The details of Taiwan's participation in the CCSBT are discussed in Chapter 8 of this book.

whose vessels conduct fishing for SBT to become members by exchange of letters.[64] Paragraph 1 of the resolution indicates that the members of the Extended Commission "shall be comprised of the Parties to the Convention and any entity or fishing entity, vessels flagged to which have caught SBT at any time in the previous three calendar years, that is admitted to membership by the Extended Commission pursuant to this Resolution."

The Extended Commission performs nearly the same function as the CCSBT, including deciding upon a total allowable catch and its allocation among the members. The CCSBT Secretariat also serves as the Secretariat of the Extended Commission. All members of the Extended Commission have equal voting rights. Theoretically, the Extended Commission meets before the Annual Commission Meeting, and its decisions are adopted in the CCSBT meeting.[65] To a certain degree, the Extended Commission, in principle, replaces the function of the CCSBT, although the latter can still adopt a resolution contrary to the former's decision. However, paragraph 4 of the resolution indicates that any decision by the CCSBT which affects the operation of the Extended Commission or the rights, obligations or status of any individual member of the Extended Commission should not be undertaken without consideration by the Extended Commission.

Although the CCSBT can adopt a resolution contrary to the decision of the Extended Commission, this circumstance should not occur because the two bodies have nearly the members; it would be contradictory for the same issue to lead to a different decision by the same members. Consequently, the CCSBT practically serves as a rubber stamp for the Extended Commission's decisions. However, for this design to work, the Extended Commission needs to include all members of the CCSBT. In other words, all CCSBT members must agree to become members of the Extended Commission, or the Extended Commission's decision might not

---

[64] Paragraph 6 of the Resolution clearly states: "Any entity or fishing entity, vessels flagged to which have caught SBT at any time in the previous three calendar years, may express its willingness to the Executive Secretary of the Commission to become a member of the Extended Commission. The Executive Secretary of the Commission, on behalf of the Commission, will conduct an Exchange of Letters with the representative of such entity or fishing entity to this effect. In so doing, the applicant shall give the Commission its firm commitment to respect the terms of the Convention and comply with such decisions of the Extended Commission as become decisions of the Commission pursuant to paragraph 4."

[65] The operation of the Extended Commission is explained in paragraphs 2, 3, 4 and 9 of the resolution.

be adopted by the commission so that the establishment of the Extended Commission lacks meaning.

As a fishing entity, Taiwan applied to become a member of the Extended Commission and the Extended Scientific Committee for CCSBT and gained that status effective 30 August 2002.[66]

7. North Pacific Anadromous Fish Commission

The NPAFC was set up on 16 February 1993 under Article VIII(1) of the Convention for the Conservation Anadromous Stocks in the North Pacific Ocean[67], signed on 11 February 1992.[68] The objective of the NPAFC is to promote the conservation of anadromous stocks in the waters of the North Pacific Ocean and its adjacent seas.[69]

The NPAFC convention does not create any route for fishing entities to accede; it is open to Canada, Japan, Russia and United States, which are major states of origin for anadromous stocks which migrate into the convention area. By a unanimous invitation from the original parties, other states may accede to the NPAFC.[70] The provisions concerning entities in the convention are in Articles IV, VI, and IX.

Under Article IV(1), the contracting parties should require states or entities which are not party to the convention to note any matter relating to the fishing activities of their nationals, residents or vessels which could affect adversely the conservation of anadromous stocks within the convention area.[71] Article IV(2) further requires the contracting parties to encourage such states or entities to adopt regulations consistent with the convention's provisions regarding fishing operations conducted by its nationals, residents or vessels.[72] The contracting parties, under Article IX(4), should "cooperate in taking action, consistent with international law

---

[66] *See* the Report of the Ninth Annual Meeting of the Commission, available at *http://www.ccsbt.org/docs/pdf/meeting_reports/ccsbt_9/report_of_ccsbt9.pdf* (visited on 19/12/2010).

[67] The full text of the convention can be found on the official NPAFC website, *http://www.npafc.org/new/publications/HandBook/English%20(page1-44).pdf* (visited on 07/01/2011).

[68] The contracting parties are Canada, Japan, Korea, Russia and the United States; further information can be found on the official NPAFC website, *http://www.npafc.org/new/about_convention.html* (visited on 07/01/2011).

[69] *Ibid.*

[70] *See* Articles XVII and XVIII, *supra note* 67.

[71] *Ibid.*

[72] *Ibid.*

and their respective domestic laws, for the prevention by any State or entity not party to this Convention of any directed fishing for, and the minimization by such State or entity of any incidental taking of, anadromous fish by nationals, residents or vessels of such State or entity in the Convention Area."[73] Article IX instructs the contracting parties to cooperate with each other and other states or entities which could hinder the conservation measures of the convention and to encourage those states or entities to adopt related laws consistent with the convention. These articles passively regulate entities rather than bringing them directly into the convention regime.

Article VI(3) continues along these lines: "The Parties shall cooperate to exchange information regarding any directed fishing for and any incidental taking of anadromous fish in the Convention Area by nationals, residents and vessels of any State or entity not party to this Convention." Article IX provides for the authority of the NPAFC.[74] Article IX(2) states that the commission can "promote the exchange of information on any activities contrary to the provisions of this Convention, especially with respect to fishing for and trafficking in anadromous fish contrary to the provisions of Article III, as well as on responsive action taken by the Parties and, as appropriate, by any State or entity not party to this Convention." Article IX(6) allows the commission to "promote the exchange of catch and effort information in respect of activities of Parties and any states or entities not party to this Convention for conducting scientific research and for coordinating the collection, exchange and analysis of scientific data regarding anadromous stocks and ecologically related species."[75] Article IX(10) is the only provision dealing with how entities may participate in the commission; it provides that the commission can invite any states or entities not party to the convention to consult with the commission on matters relating to the conservation of anadromous stocks and ecologically related species in the convention area.[76]

Although entities may consult on matters relating to the conservation and management of anadromous stocks with the commission, they may not accede to the convention or become members of the NPAFC. The regulations concerning entities in the NPAFC Convention passively request that the commission cooperate with entities in order to acquire fisheries information or prevent such entities from violating the conservation and management measures adopted by the NPAFC. However,

---

[73] *Ibid.*

[74] *Ibid.*

[75] *Ibid.*

[76] *Ibid.*

those entities which are neither contracting parties nor members have no obligation to comply with the NPAFC's regulations. Fishing entities which are invited to attend the meetings and discuss some particular issues and agree to comply with certain stipulations or resolutions of the NPAFC can revoke the agreement at any time. Hence, the NPAFC convention's position concerning entities does not provide a good way either to establish or to stabilise relations with fishing entities. A better way for the NPAFC to prevent fishing entities from breaking its conservation and management measures would be to bring them into its legal regime, in other words, to create a route for fishing entities to become contracting parties or members of the NPAFC required to comply with its resolutions. The NPAFC should consider amending its convention, perhaps using the CCSBT's experience as a reference.

Taiwan, as a fishing entity in other RFMOs, has been invited to be an observer and join in discussions at the commission's annual meetings since 2005. However, it is not been identified as a fishing entity but has merely used the term "Taiwan" to participate.[77]

8. Indian Ocean Tuna Commission

The FAO established the IOTC on 27 March 1996 to manage tuna and tuna-like species in the Indian Ocean and adjacent seas and promote cooperation among members to ensure the conservation and optimum utilisation of stocks and encourage the sustainable development of fisheries based on such stocks.[78] The IOTC falls within the framework of FAO and, thus, also within the framework of the UN. This position limits the membership in the IOTC to only FAO members and associate members, UN members and specialized agencies and members of the International Atomic Energy Agency.[79] The Agreement for the Establishment

---

[77] The Annual Meeting Reports from 1993 are available at *http://www.npafc.org/new/pub_annualreport.html* (visited on 07/01/2011).

[78] Further information can be found on the IOTC website, *http://www.iotc.org/English/index.php* (visited on 30/12/2010).

[79] The IOTC members are Australia, Belize, China, Comoros, Eritrea, the European Union, France, Guinea, India, Indonesia, Iran, Japan, Kenya, Korea, Madagascar, Malaysia, Mauritius, the Sultanate of Oman, Pakistan, Philippines, Seychelles, Sierra Leone, Sri Lanka, Sudan, Tanzania, Thailand, the United Kingdom and Vanuatu. The cooperating non-contracting parties are the Maldives, Senegal, South Africa and Uruguay. Further information can be found at *http://www.iotc.org/English/info/comstruct.php* (visited on 17/01/2011).

of the Indian Ocean Tuna Commission[80] did not create any approach for fishing entities to accede to or join the IOTC. One provision in the agreement concerning fishing entities is Article X(4), which requires IOTC members to cooperate and exchange information on stocks covered by the convention with non-member fishing entities whose vessels fish in the convention area. As well, Article XI(1) provides that the IOTC should make effort to obtain fishing statistics from fishing states or entities which are not members.

At the Fourth Meeting of the IOTC in 1999, Resolution 99/04 on the Status of Cooperating Non-Contracting Parties was adopted, defining "cooperating non-contracting parties" as "[a]ny non-Contracting Party that voluntarily ensures that vessels flying its flag fish in a manner which is in conformity with the conservation measures adopted by IOTC be defined as a Non-Contracting Cooperating Party."[81] The resolution required the IOTC secretary to encourage non-contracting parties to become contracting parties or non-contracting cooperating parties.[82] In 2003, Resolution 99/04 was replaced by Resolution 03/02 on Criteria for Attaining the Status of Co-Operating Non-contracting Party. It stipulates that co-operating non-contracting parties' status should be reviewed and renewed annually and that any non-contracting parties requesting the status of co-operating non-contracting party should provide: "a) where available, data on its historical fisheries in the IOTC Area, including nominal catches, number/type of vessels, name of fishing vessels, fishing effort and fishing areas; b) all the data that Contracting Parties have to submit to IOTC based on the resolutions adopted by IOTC; c) details on current fishing presence in the IOTC Area, number of vessels and vessel characteristics and; d) information on any research programmes it may have conducted in the IOTC Area and the information and the results of this research."[83] Although Resolution 03/02 does not define the term "cooperating non-contracting parties," it in principle includes fishing entities.

However, the critical problem that the IOTC faces concerning fishing entities is Taiwan, which is the major catcher of tuna in the Indian Ocean.

---

[80] The full text of the agreement can be found on the IOTC website, *http://www.iotc.org/files/proceedings/misc/ComReportsTexts/IOTC%20Agreement.pdf* (visited on 18/02/2011).

[81] The full text of the resolution can be found at: *http://www.iotc.org/files/proceedings/misc/ComReportsTexts/resolutions_E.pdf* (visited on 11/01/2011).

[82] *Ibid.*

[83] *Ibid.*

The commission's conservation and management efforts cannot be effective without Taiwan's participation.[84] However, the IOTC is the only regional fisheries organisation that excludes Taiwan from participating. As the IOTC falls within the framework of the FAO and UN, it is involved in the dispute over recognition between China and Taiwan.[85] To solve this problem, the seventh meeting of the commission in 2002 explored a legal route for Taiwan to be brought into the IOTC as a cooperative non-member entity.[86] So far, however, the commission has failed to create or find a way for Taiwan. Taiwan has participated as an invited expert, a status which recognises individuals rather than government and lacks the rights and obligations of members.[87]

9. South East Atlantic Fisheries Organisation

In 1995, Namibia suggested establishing a regional fisheries management organisation to conserve and manage fishery resources in the South-east Atlantic Ocean.[88] Through nine preparatory meetings between 1997 and 2001, the Convention on the Conservation and Management of Fisheries Resources in the South East Atlantic Ocean[89] was negotiated. It was signed in April 2001[90] and entered into force in April 2003, leading to the establishment of the SEAFO.[91] The convention was the first to create a regional management organisation after the adoption of the UNFSA. The convention's objective is to ensure the long-term conservation and sustainable use of the fishery resources in the convention area.

---

[84] *See* W.R. Edeson, "An International Legal Extravaganza in the Indian Ocean: Placing the Indian Ocean Tuna Commission outside the Framework of FAO," *The International Journal of Marine and Coastal Law*, Vol. 22, No.4(2007), p. 486; William Edeson, "Some Future Directions for Fishing Entities in Certain Regional Fisheries Management Bodies," *Ocean Development and International Law*, Vol. 37(2006), p. 252.

[85] Regarding the dispute between China and Taiwan, *see* Chapter 7 of this book.

[86] *See* the 8th Meeting of the Commission, available at *http://www.iotc.org/files/proceedings/2003/s/IOTC-2003-S08-R[EN].pdf*

[87] William Edeson, *supra note* 84, p. 261.

[88] *See* the website of SEAFO, *http://www.seafo.org/welcome.htm* (visited on 11/01/2011).

[89] The full text of the convention can be found on SEAFO website, *http://www.seafo.org/welcome.htm* (visited on 18/02/2011).

[90] *Ibid.*

[91] *Ibid.* The parties to the SEAFO are Angola, the European Union, Japan, Namibia, Norway and South Africa.

Article 1(i) defines "fishing entity" as "any fishing entity referred to in article 1 paragraph 3 of the 1995 agreement." However, the convention does not contain any provisions for a fishing entity allowing to accede. Article 26 states that the convention is open to accession by coastal states and all other states and regional economic integration organisations whose vessels fish in the convention area.[92] Like the NPAFC, the SEAFO regulations concerning fishing entities all require contracting parties to cooperate with fishing entities in order to prevent any diminishment of the effectiveness of the conservation and management measures. Article 6(10) requires the commission to "draw the attention of any State or fishing entity which is a non-party to this Convention to any activity which in the opinion of the Commission affects implementation of the objective of this Convention."[93] Article 22(4) instructs contracting parties to cooperate with fishing entities that have fishing vessels in the convention area in implementing conservation and management measures.[94] This article further states that "such fishing entities shall enjoy benefits from participation in the fishery commensurate with their commitment to comply with conservation and management measures in respect of the stocks"[95]—the same provision as in Article 17(3) of the UNFSA. However, the convention does not further explain what the commensurate benefits and obligations of fishing entities are.

At present, the Convention of the SEAFO excludes fishing entities from membership, nor has any fishing entity joined the SEAFO.

10. Conclusion

RFMOs are usually established because coastal states and states with vessels fishing in certain high seas areas see the necessity for regional organisations to conserve and manage fisheries resources. The strategy that the RFMOs adopt is to invite all related states, especially those whose vessels fish in the convention areas, to enter the RFMO's regime. Those states either become contracting parties or members of the RFMOs, facilitating the effective implementation of the RFMO's conservation and management measures. As the amount of regional fisheries resources is limited, some states catch more, decreasing the amount that other states can catch. Consequently, the most important RFMO conservation and

---

[92] *Ibid.*

[93] *Ibid.*

[94] *Ibid.*

[95] *Ibid.*

management measure is to use scientific statistics provided by each member to calculate a TAC in order to avoid over-fishing. RFMOs allocate quotas to each member according to the TAC. RFMOs need all states whose vessels fish in their convention areas to cooperate so that the scientific statistics, TAC and allocated quotas can be calculated accurately. Therefore, brining all actors, even non-traditional actors, into the RFMOs' conservation and management regimes is key to effectively carrying out the conservation and management measures. Thus, most of the RFMOs discussed bring fishing entities, non-traditional actors, into their legal system.

If fishing entities are excluded from RFMO regimes, they do not have the obligation to comply with the conservation and management measures adopted by the RFMOs. First, the statistical information that RFMOs gather might be inaccurate, which would influence calculation of the TAC. If the TAC has a huge discrepancy with reality, over-fishing might occur even if each member strictly complies with the allocated quotas. Furthermore, even if calculation of the TAC is not influenced by inaccurate statistics and each member catches only their allocated quota, fishing entities not bound by the RFMO regulations might conduct unlimited fishing within the RFMOs' convention areas, exceeding the TAC and making the efforts of the RFMO members in vain. Consequently, binding all actors within an RFMO's convention area to its legal system is crucial to effectively implement the RFMO's conservation and management measures. Therefore, the WCPFC, IATTC, and ISC endow fishing entities with rights equal to those of states which become members, and the ICCAT and CCSBT allow a route for fishing entities to participate and be bound by the organisations' resolutions. In addition, members of the IOTC tried to amend its convention to allow fishing entities to participate. The NPAFC and SEAFO have not opened a door for fishing entities to join possibly because their conservation and management measures are not affected by fishing entities. Once those non-traditional actors become a factor in decreasing the effectiveness of conservation and management, these two RFMOs can be expected to create a route for fishing entities to participate and be bound by their legal regimes.

PART II:

ENFORCEMENT IN HIGH SEAS FISHERIES BY FISHING ENTITIES

CHAPTER FOUR

FISHING ENTITIES AS EQUIVALENT TO FLAG STATES IN HIGH SEAS ENFORCEMENT

1. Introduction

The high seas encompass all parts of the sea not under the jurisdiction of any state. However, the extent of the high seas is not constant but changes over time. Pursuant to Article 1 of the 1958 Convention on the High Seas[1], the term "high seas" is defined as "all parts of the sea that are not included in the territorial sea or in the internal waters of a State." Due to the extension of states' jurisdiction over the ocean, the concept of exclusive economic zones (EEZ) was built into the 1982 UNCLOS, reducing the scope of the high seas to "all parts of the sea that are not included in the EEZ, in the territorial sea or in the internal waters of a State, or in the archipelagic waters of an archipelagic State."[2] In the seventeenth century, Hugo Grotius argued that, as the resources of the seas were limitless and inexhaustible, they could not be occupied by or subject to anyone.[3] The modern freedom of high seas stems from Grotius' contention.

Pursuant to Article 2 of the 1958 Convention on the High Seas, the freedom of high seas includes the freedom of navigation and fishing and the freedom to lay submarine cables and pipelines and to fly over the high seas. In addition to these four freedoms of the high seas, the 1982 UNCLOS adds two more: the freedom to construct artificial islands and other installations permitted under international law and freedom of scientific research. These six freedoms of the high seas lead to a corollary: the principle of the exclusivity of flag state jurisdiction. As the existence of fishing entities relies on their specific function of fishing, this chapter

[1] United Nations, *Treaty Series* (Vol. 450), p. 11; the full text is also available at *http://untreaty.un.org/ilc/texts/instruments/english/conventions/8_1_1958_high_seas.pdf* (visited on 03/05/2011).

[2] *See* Article 86 of the UN Convention on the Law of the Sea.

[3] Hugo Grotius, *The Freedom of the Seas* (trs. Ralph Van Deman Magoffin, New York: Oxford University Press, 1916), pp. 27–28.

does not discuss all of the six freedoms but focuses on the role of fishing entities in the freedom of fishing and their exercise of the exclusivity of flag state jurisdiction. This chapter shows that fishing entities enjoy the freedom of fishing and the attendant obligations, and on the basis of the freedom of fishing and its limitations, examines fishing entities' exclusive jurisdiction and enforcement on the high seas.

2. The Freedom of Fishing on the High Seas

The freedom of fishing means that the fishery resources on the high seas are open for all states to fish and catch. This principle originated from the reasoning that, although the high seas are the common property of all (*res communis omnium*), the natural resources within the high seas are not subject to anyone specific (*res nullius*). Thus, the nationals of all states have the right to engage in fishing, and their vessels are free to catch and possess those resources on the high seas.[4]

However, it must be noted that the freedom of fishing is given not to individuals but to all states, both coastal and landlocked.[5] Article 2 of the 1958 Convention on the High Seas states that the freedom of high seas extends to both coastal and non-coastal states. Article 87 of the 1982 UNCLOS reasserts this rule and further indicates that freedom of high seas is for both coastal and "land-locked" states, replacing the term "non-coastal" states. The principle that states are the main subject of international law is not the only reason that the freedom of fishing is given to states rather than individuals. Practically, as a result of *res communis omnium*, the special status of high seas, Article 2 of the 1958 Convention on the High Seas and Article 89 of the 1982 UNCLOS stipulate that no state may subject any part of the high seas to its sovereignty. Consequently, no states can control or rule the high seas. Therefore, the effective procedure to maintain order in the activity of fishing on the high seas is for states to control their vessels which engage in fishing on the high seas.

In the international law of the sea, fishing entities are defined by their specific function of fishing. They cannot be regarded as the same as normal states; however, they absolutely possess the capacity for fishing. Although fishing entities are different from normal states,[6] the operation of

---

[4] *See* Article 2 of 1958 Convention on the High Seas and articles 87 and 116 of UNCLOS.

[5] Rosemary Gail Rayfuse, *Non-flag State Enforcement in High Seas Fisheries* (Leiden: Martinus Nijhoff Publishers, 2004), p. 30.

[6] In this chapter, we do not deal with how and why fishing entities become entities, instead of normal states because there are complicated reasons which vary case by

their vessels which engage in fishing on the sea is the same as those of states. As owners of vessels, their nationals register with the authorities of fishing entities, placing fishing entities in the position of *de facto* flag states. The vessels of fishing entities should comply with the rules and domestic laws stipulated by fishing entities. Like flag states, fishing entities as *de facto* flag states are under the obligation to restrict the fishing activities of their nationals on the high seas and to control their vessels so as not to violate the international law of the sea.[7]

Despite having the same duties as flag states, fishing entities, as *de facto* flag states, are not the subject of the 1958 Convention on the High Seas and the 1982 UNCLOS. Neither convention mentions the term "fishing entities." Article 31 of the 1958 Convention on the High Seas states that it may be signed by all UN member states and specialised agencies and by any other state invited by the UN General Assembly to become a party to the convention. Article 33 opens accession to the convention to states belonging to any of the categories mentioned in Article 31. The 1958 Convention on the High Seas does not create opportunities for fishing entities to sign or accede.[8] Articles 305 and 307 of the UNCLOS also do not provide a gateway for fishing entities to be the subject of the convention.[9] However, lacking the opportunity to be the subject of those two conventions does not prevent fishing entities from enjoying the freedom of fishing on the high seas. The 1958 Convention on the High Seas and most related fishing rules in UNCLOS are not newly developed or created; rather, they codify customary international law which has long existed and is viewed by international society as "evidence of a general practice accepted as law."[10] The freedom of fishing is

---

case. Neither do we discuss the extent of how the concept of fishing entities differs from that of the state. At present, international law does not present an objective standard which can be judged.

[7] This assumes that fishing entities are subject to the law of the sea. The cross-references regarding fishing entities' international legal personality are shown in Chapter 2 of this book.

[8] The concept of fishing entities had not developed at that time. The international document which first referenced fishing entities is a paper entitled "International Fishery Bodies: Considerations for High Seas Management," from the FAO Technical Consultation on High Seas Fishing. This paper is discussed in Chapter 1 of this book.

[9] The details of fishing entities' status in UNCLOS are discussed in Chapter 2 of this book.

[10] *See* Article 38 of the Statute of the International Court of Justice. Also *see* Ian Brownlie, *Principles of Public International Law*, 7th ed. (New York: Oxford

regarded as customary international law.[11] Theoretically, the actors of international society should comply with customary international law and could be endowed with some rights from customary law. Fishing entities are no exception. Therefore, although they are not the subject of the 1958 Convention on the High Seas or UNCLOS, fishing entities are granted the right of freedom of fishing by international customary law. In other words, the practice of vessels of fishing entities exercising the freedom of fishing by engaging in fishing on high seas is not regarded as violating international law.

In contrast, the term "fishing entity" was included in the UNFSA and the related regulations of RFMOs. Pursuant to Article 1(3) of the UNFSA, it applies to fishing entities whose vessels fish on high seas. This article implies that it is necessary to bring fishing entities into the legal system in order to achieve the conservation and sustainable use of straddling and highly migratory fish stocks. Meanwhile, it can be reasoned that vessels of these fishing entities have been engaging in fishing on high seas, so their fishing activities must be regulated. Similarly, the regulations of some RFMOs clearly allow fishing entities to become members. As members of or participants in RFMOs, fishing entities assume obligations imposed by the RFMOs' conservation and management measures, including allocation of catch quota and detailed stipulations for vessels fishing on the high seas. The regulations of the UNFSA and RFMOs imply that fishing entities invited to become the members of or participants in RFMOs possess the right to fish on high seas. On one hand, this practice seems to contradict traditional international law of the sea which grants that the freedom of fishing only to states. On the other, it can be regarded as recognising fishing entities' conduct of fishing on the high seas so as to imply that, in addition to states, fishing entities can also enjoy the freedom of fishing. However, the purpose of RFMOs' conservation and management measures is to protect and sustain fisheries resources on the high seas by reducing the total catch amount per year. Arguably, given the aim to reduce the catch amount of certain fish stocks on the high seas, it is not in RFMOs' interest to increase the vessels fishing those stocks on the high seas by endowing with rights new legal actors not originally permitted to fish on the high seas. The reasonable inference is that the fishing entities which can become members of or participants in RFMOs' conservation and management measures are recognised as also originally enjoying the

---

University Press, 2008), p. 8; *see* also the Preamble of the 1958 Convention on the High Seas.

[11] R.R. Churchill and A.V. Lowe, *The Law of the Sea*, 3rd ed. (Manchester: Manchester University Press, 1999), p. 203.

freedom of fishing. Otherwise, it would be unnecessary to bring this kind of fishing entity into the legal system of conservation and management measures, limit their catching amount and require their cooperation. Therefore, it can be deduced that fishing entities within RFMOs' system already possess the freedom of fishing on the high seas, so the related rules of conservation and management measures adopted by RFMOs may reasonably be imposed on them.

3. The Limitations of the Freedom of Fishing

Although the freedom of fishing brings many benefits and convenience for humans, it is still accompanied by some limitations. Article 2 of the 1958 Convention on the High Seas states that, when exercising freedoms of the high seas, states should have reasonable regard for the interests of other states. Article 87(2) of UNCLOS reads that "[t]hese freedoms shall be exercised by all States with due regard for the interests of other States in their exercise of the freedom of the high seas, and also with due regard for the rights under this Convention with respect to activities in the Area." These articles remind all states to consider other states' rights and freedoms as they also exercise these rights. The purpose of this limitation is to hinder the exercise of these freedoms by each state and to balance the benefits of using the high seas for each state.

In addition, the freedom of fishing is largely subject to the measures for conservation and management on high seas fisheries resources. International society acknowledges that the high seas fisheries resources are not limitless and inexhaustible, as Grotius thought. Therefore, regulations concerning the conservation and management of these resources have been established. The 1958 Convention on Fishing and Conservation of the Living Resources of the High Seas[12] was adopted and entered into force on 20 March 1966. Article 1(2) of this convention states that all states have the duty to adopt or cooperate with other states in adopting necessary measures for the conservation of the living resources of the high seas. Article 4 further indicates that, if the nationals of two or more states engage in fishing the same stocks of fish or other living marine resources on the high seas, these states shall negotiate to adopt necessary measures for the conservation of the living resources affected. The 1982

---

[12] United Nations, *Treaty Series* (Vol. 559), p. 285; the full text is also available at *http://untreaty.un.org/ilc/texts/instruments/english/conventions/8_1_1958_fishing.pdf* (visited on 03/05/2011).

UNCLOS has provisions similar to the 1958 convention.[13] Article 117 of the UNCLOS reads: "All States have the duty to take, or to cooperate with other States in taking, such measures for their respective nationals as may be necessary for the conservation of the living resources of the high seas." Article 118 stipulates that states whose nationals exploit either the same or different living resources in the same area should undertake the measures necessary for the conservation of these living resources and cooperate in establishing subregional or regional fisheries organisations. Article 116 further indicates that the freedom of fishing is limited by measures adopted by coastal states to conserve and manage straddling and highly migratory fish stocks in EEZs.

Moreover, in 1991 the UN adopted a Resolution on Large-scale Pelagic Drift-net Fishing and its Impact on the Living Marine Resources of the World's Oceans and Seas[14], which requires all members of international community to cease large-scale pelagic drift-net fishing on the high seas from the end of 1992. The Agreement to Promote

---

[13] However, some of provisions of UNCLOS are more specific. For example, Article 119 reads:

> "1. In determining the allowable catch and establishing other conservation measures for the living resources in the high seas, States shall:
>
> (a) take measures which are designed, on the best scientific evidence available to the States concerned, to maintain or restore populations of harvested species at levels which can produce the maximum sustainable yield, as qualified by relevant environmental and economic factors, including the special requirements of developing States, and taking into account fishing patterns, the interdependence of stocks and any generally recommended international minimum standards, whether subregional, regional or global;
>
> (b) take into consideration the effects on species associated with or dependent upon harvested species with a view to maintaining or restoring populations of such associated or dependent species above levels at which their reproduction may become seriously threatened.
>
> 2. Available scientific information, catch and fishing effort statistics, and other data relevant to the conservation of fish stocks shall be contributed and exchanged on a regular basis through competent international organizations, whether subregional, regional or global, where appropriate and with participation by all States concerned.
>
> 3. States concerned shall ensure that conservation measures and their implementation do not discriminate in form or in fact against the fishermen of any State."

[14] *A/RES/46/215*. The full text can be found at *http://daccess-dds-ny.un.org/doc/RESOLUTION/GEN/NR0/583/03/IMG/NR058303.pdf?OpenElement* (visited on 05/05/2011).

Compliance with International Conservation and Management Measures by Fishing Vessels on the High Seas and the Code of Conduct for Responsible Fisheries, adopted by FAO in 1993 and 1995 respectively, require flag states to monitor their vessels which engage in fishing on the high seas to ensure that these vessels do not violate relevant conservation and management measures.[15] Additionally, the UNFSA plays a significant role in encouraging states to cooperate in adopting conservation and management measures for straddling and highly migratory fish stocks in the high seas through subregional or regional fisheries management organisations.

As the concept of fisheries resources conservation and management have gained international concern, related measures, including many treaties, have largely restricted the traditional freedom of fishing. While exercising the freedom of fishing like states, fishing entities, too, are bound by these limitations. In particular, the UNFSA and other treaties creating RFMOs seek to bring fishing entities under their conservation and management measures on high seas. Fishing entities thus cannot escape responsibility for high seas fisheries conservation and management while enjoying the freedom of fishing.

---

[15] Article 3(3) of the Agreement to Promote Compliance with International Conservation and Management Measures by Fishing Vessels on the High Seas states: "No Party shall authorize any fishing vessel entitled to fly its flag to be used for fishing on the high seas unless the Party is satisfied that it is able, taking into account the links that exist between it and the fishing vessel concerned, to exercise effectively its responsibilities under this Agreement in respect of that fishing vessel." Article 3(6) reads: "Each Party shall ensure that all fishing vessels entitled to fly its flag … are marked in such a way that they can be readily identified in accordance with generally accepted standards." Article 3(7) further states: "Each Party shall ensure that each fishing vessel entitled to fly its flag shall provide it with such information on its operations as may be necessary to enable the Party to fulfil its obligations under this Agreement, including in particular information pertaining to the area of its fishing operations and to its catches and landings." Article 6.10 of the Code of Conduct for Responsible Fisheries states: "States should ensure compliance with and enforcement of conservation and management measures and establish effective mechanisms, as appropriate, to monitor and control the activities of fishing vessels and fishing support vessels." Article 7.1.7 also reads: "States should establish…effective mechanisms for fisheries monitoring, surveillance, control and enforcement to ensure compliance with their conservation and management measures." Article 7.7.3 reiterates this, stipulating that "[s]tates … should implement effective fisheries monitoring, control, surveillance and law enforcement measures including, where appropriate, observer programmes, inspection schemes and vessel monitoring systems."

4. The Exclusivity of Flag State Jurisdiction and Enforcement on High Seas

4.1 Exclusive Jurisdiction

The principle of the freedom of high seas allows every state to use the high seas. As no state can demand or exercise sovereignty over the high seas, vessels from all states are equal and cannot be controlled and subjected to the jurisdiction of other states, *par in parem non habet imperium*. As a corollary, each state thus has the obligation not to interfere with vessels of other flag states.[16] Consequently, the exclusivity of flag state jurisdiction can be applied when vessels fish on the high seas.[17]

This application is qualified by the 1927 *Lotus* case[18], in which the Permanent Court of International Justice decided that an officer responsible for a collision could be prosecuted by Turkey, as well as France, because the collision between French and Turkish ships which caused a loss of life on the Turkish ship had taken place on the Turkish ship as well as the French ship. This decision was much criticised and reversed by the 1952 International Convention for the Unification of Certain Rules Relating to Penal Jurisdiction in Matters of Collision or other Incidents of Navigation.[19] Under it, proceedings concerning criminal or disciplinary responsibility for the masters or any persons in ships involved in collisions or navigation incidents while sea-going may be performed only by the judicial or administrative authorities of the flag states.[20] The rule of the 1952 Brussels Convention was adopted in the 1958 Convention on the High Seas and the UNCLOS. Pursuant to Article 11 of the 1958 High Seas Convention and Article 97 of UNCLOS, penal and disciplinary proceedings in cases of collision or other navigational

---

[16] Robert C.F. Reuland, "Interference with Non-national Ships on the High Seas: Peacetime Exceptions to the Exclusivity of Flag State Jurisdiction," *Vanderbilt Journal of Transnational Law*, Vol. 22(1989), pp. 1164–1165.

[17] *Ibid.*

[18] *See Publications of the Permanent Court of International Justice* (Series A.-No 10). Also *see* the official website of the International Court of Justice: *http://www.icj-cij.org/pcij/series-a.php?p1=9&p2=1* (visited on 04/11/2011).

[19] The convention was adopted and signed in Brussels on 10 May 1952 and entered into force on 20 November 1955. *See* United Nations, *Treaty Series* (Vol. 439), p. 235; the full text is available at *http://treaties.un.org/doc/publication/unts/volume%20439/volume-439-i-6332-english.pdf* (visited on 03/10/2011).

[20] *See* Article 1 of the 1952 Convention.

incidents are held by the authorities of that state in whose ship the defendants served or of the state of which they are nationals. A state retains jurisdiction over its nationals wherever they might be, but in the case of concurrent jurisdiction, the flag states have primacy.[21]

The jurisdiction of the flag state on the high seas is not only a right but also an obligation. Apart from the duties of the master, crew and passengers in a ship flying its flag to render assistance to ships in distress and to enforce legislation dealing with those matters,[22] each flag state has the obligation to create laws making it an offence for their ships fishing on the high seas to break or injure submarine cables or pipelines under the high seas and to offer compensation for such acts.[23] Furthermore, pursuant to Article 5[24] of the 1958 Convention on the High Seas and Article 91[25] of the UNCLOS, there must be a genuine link between ships fishing on the high seas and their flag states. Neither convention defines "genuine link." Although the rule regarding a genuine link is included in customary international law and codified in 1958 convention, there is still no agreement on what constitutes the genuine link that customary international law requires.[26] Article 5 of

---

[21] R.R. Churchill and A.V. Lowe, *supra note* 11, p. 209.

[22] *See* articles 10 and 11 of the 1958 Convention on the High Seas and articles 94 and 98 of the 1982 UNCLOS.

[23] For the details, *see* Articles 27–29 of the 1958 High Seas Convention and Articles 113–115 of the 1982 UNCLOS.

[24] Article 5 reads: "Each State shall fix the conditions for the grant of its nationality to ships, for the registration of ships in its territory, and for the right to fly its flag. Ships have the nationality of the State whose flag they are entitled to fly. There must exist a genuine link between the State and the ship; in particular, the State must effectively exercise its jurisdiction and control in administrative, technical and social matters over ships flying its flag."

[25] Article 91 states: "Every State shall fix the conditions for the grant of its nationality to ships, for the registration of ships in its territory, and for the right to fly its flag. Ships have the nationality of the State whose flag they are entitled to fly. There must exist a genuine link between the State and the ship."

[26] D.J. Harris presents two opposite examples to show that it is unclear whether the "genuine link" requirement is a part of customary international law. The *travaux preparatoires* of the 1986 UN Convention on Conditions for Registration of Ships shows widespread support for it; however, Liberia asserted that "any limitation on the rights of states to determine the conditions under which vessels should be accepted on national shipping registers would be contrary to customary international law." *See* D.J. Harris, *Cases and Materials on International Law*, 7th ed. (London: Sweet & Maxwell, 2010), p. 368. The 1986 Registration Convention stipulated many conditions which could be considered to determine whether there is a genuine link; however, this convention has never come into force. The full text of the 1986 Registration Convention can be found on the website of Admiralty and

High Seas Convention indicates that "the State must effectively exercise its jurisdiction and control in administrative, technical and social matters over ships flying its flag," as does Article 94 of the UNCLOS. Although it is unclear in Article 5 whether a state's exercise of such jurisdiction is a condition or consequence of a genuine link,[27] the effective exercise of jurisdiction and control likely could serve as evidence of a genuine link between flag states and fishing vessels. In light of Article 94 of UNCLOS, states have the obligation to ensure that their vessels fishing on high seas are under their effective control by maintaining a register and assuming jurisdiction in internal law over all vessels fishing on the high seas and their masters, officers and crew in administrative, technical and social matters concerning the ship. States must also ensure the safety of their vessels fishing on high seas, *inter alia*, checking (a) the construction, equipment and seaworthiness of ships; (b) the manning of ships, labour conditions and the training of crews; and (c) the use of signals, the maintenance of communications and the prevention of collisions.

The UNFSA also lists flag states' duties regarding their vessels fishing on the high seas which establish a genuine link between flag states and vessels, which include: ensuring that their vessels comply with subregional and regional conservation and management measures; authorising vessels to fish on the high seas only when they can effectively fulfil the responsibilities for such vessels under UNCLOS and the agreement; and making sure that the measures they impose on their vessels are compatible with the subregional, regional and global systems of monitoring, control and surveillance.[28]

Maritime Law Guide, *http://www.admiraltylawguide.com/conven/registration1986.html* (visited on 04/11/2011).

[27] *See* Rayfuse, *supra note* 5, p. 26.

[28] *See* Article 18 of the UNFSA. It stipulates that flag states should take measures, including:

> "(a) control of such vessels on the high seas by means of fishing licences, authorizations or permits, in accordance with any applicable procedures agreed at the subregional, regional or global level;
> (b) establishment of regulations:
> (i) to apply terms and conditions to the licence, authorization or permit sufficient to fulfil any subregional, regional or global obligations of the flag State;
> (ii) to prohibit fishing on the high seas by vessels which are not duly licensed or authorized to fish, or fishing on the high seas by vessels otherwise than in accordance with the terms and conditions of a licence, authorization or permit;

(iii) to require vessels fishing on the high seas to carry the licence, authorization or permit on board at all times and to produce it on demand for inspection by a duly authorized person; and
(iv) to ensure that vessels flying its flag do not conduct unauthorized fishing within areas under the national jurisdiction of other States;
(c) establishment of a national record of fishing vessels authorized to fish on the high seas and provision of access to the information contained in that record on request by directly interested States, taking into account any national laws of the flag State regarding the release of such information;
(d) requirements for marking of fishing vessels and fishing gear for identification in accordance with uniform and internationally recognizable vessel and gear marking systems, such as the Food and Agriculture Organization of the United Nations Standard Specifications for the Marking and Identification of Fishing Vessels;
(e) requirements for recording and timely reporting of vessel position, catch of target and non-target species, fishing effort and other relevant fisheries data in accordance with subregional, regional and global standards for collection of such data;
(f) requirements for verifying the catch of target and non-target species through such means as observer programmes, inspection schemes, unloading reports, supervision of transshipment and monitoring of landed catches and market statistics;
(g) monitoring, control and surveillance of such vessels, their fishing operations and related activities by, inter alia:
(i) the implementation of national inspection schemes and subregional and regional schemes for cooperation in enforcement pursuant to articles 21 and 22, including requirements for such vessels to permit access by duly authorized inspectors from other States;
(ii) the implementation of national observer programmes and subregional and regional observer programmes in which the flag State is a participant, including requirements for such vessels to permit access by observers from other States to carry out the functions agreed under the programmes; and
(iii) the development and implementation of vessel monitoring systems, including, as appropriate, satellite transmitter systems, in accordance with any national programmes and those which have been subregionally, regionally or globally agreed among the States concerned;
(h) regulation of transshipment on the high seas to ensure that the effectiveness of conservation and management measures is not undermined; and
(i) regulation of fishing activities to ensure compliance with subregional, regional or global measures, including those aimed at minimizing catches of non-target species."

While stipulating related conservation and management measures concerning straddling and highly migratory fish stocks, RFMOs bring fishing entities into their legal system by establishing certain relationships with fishing entities, such as membership, observer or invited expert. As a member of a RFMO, fishing entities' have to comply with conservation and management measures, such as allocations of allowable catch or levels of fishing effort for their vessels fish on the high seas. This system implies that, as fishing entities are recognised as possessing the right to fish these stocks on the high seas, they need to obey the related measures. As fishing entities are admitted to legally fish on the high seas, there is no reason to think that they can escape the duties imposed on flag states. Therefore, the duties of flag states under the 1995 agreement also apply to fishing entities that are members of RFMOs or regarded as possessing the right to fish on the high seas.

The 1958 High Seas Convention and UNCLOS do not clarify whether the stipulations concerning the jurisdiction of the flag state can apply to fishing entities, *i.e.* whether fishing entities can exert jurisdiction over the vessels flying their flags. The basis of flag states' jurisdiction on the high seas stems from flag states' duties to effectively control their vessels and prevent their vessels from violating international conventions and customs. However, the duties of flag states on high seas are based on the freedom of fishing. As mentioned, fishing entities can fish on the high seas based on RFMOs regulations. Nevertheless, the conservation and management measures adopted by RFMOs further restrict the scope of the freedom of fishing on the high seas. As the purpose of RFMOs' measures is to conserve and sustainably use fisheries resources by reducing the total amount of catch per year, fishing entities must be under the same obligations as flag states to control vessels flying their flags and fishing on the high seas. Thus, fishing entities must also possess jurisdiction over their vessels, as flag states do under the 1958 Convention on the High Seas and UNCLOS. In other words, without jurisdiction over vessels flying their flag, fishing entities' right to engage in fishing on the high seas would not be meaningful, as fishing can only be done through the means of vessels.

4.2 Enforcement on the High Seas

To fulfil the obligations to ensure that vessels fishing on the high seas comply with international conventions, flag states generally possess exclusive jurisdiction over their vessels. Flag states, therefore, can conduct enforcement on these vessels in accordance with domestic regulations.[29] Due to the limitation of the environment at sea, enforcements on the high seas usually include stop, approach, boarding, search and arrest.[30]

Under traditional international law of the sea, both the 1958 and 1982 conventions rule that, unless a ship on high seas is reasonably suspected of engaging in piracy, the slave trade or unauthorised broadcasting, or of lacking nationality, non-flag states' warships, military aircraft or any other ship or aircraft clearly marked and identifiable as in government service cannot board and inspect a foreign ship.[31] A non-flag state may board and inspect a foreign ship which, though flying a foreign flag or refusing to show its flag, is suspected of having the same nationality as the enforcing ship.[32] It is not debated whether an authorised unit, including a warship or a ship in government service, may board, inspect and arrest ships which are fishing on the high seas and flying the same flag as the authorised unit following its own domestic law, although this practice is not explicitly mentioned in international conventions which usually only stipulate the conditions for non-flag state enforcement.

To control vessels on the high seas flying the flags of fishing entities and to ensure those vessels comply with international conventions, including RFMOs' conservation and management measures, fishing entities can be presumed to be able to conduct enforcements on high seas according to their internal regulations. Their authorised ships can impose enforcements on the vessels flying their flags on the high seas, as flag states do under the 1958 Convention on the High Seas, UNCLOS and Fish Stock Agreement.

5. Conclusion

Under the traditional international law of the sea, only states may enjoy the freedom of fishing on the high seas. Before the 1990s, the concept of fishing entities did not exist. Since the UNFSA formally introduced the term fishing entity in its stipulations, fishing entities primarily have been recognised in agreements creating RFMOs and resolutions issued by RFMOs. The concept of the fishing entity thus has been established in international fishery law. Reviewing UNFSA regulations and agreements

---

[29] *See* Reuland, *supra note* 16, pp. 1163-1164; R.R. Churchill and A.V. Lowe, *supra note* 11, p. 208.

[30] Reuland, *supra note* 16, pp. 1169–1176.

[31] *See* Article 22 of the 1958 Convention on the High Seas and Article 110 of 1982 UNCLOS.

[32] *Ibid.*

creating RFMOs, it can be seen that fishing entities, as well as states, are able to enjoy the freedom of fishing on the high seas. As a corollary to the freedom of fishing, fishing entities are bound by the accompanying limitations stipulated in the 1958 Convention on the High Seas, the 1958 Convention on Fishing and Conservation of the Living Resources of the High Seas and UNCLOS. Fishing entities also have exclusive jurisdiction like a flag state and can conduct enforcement against their vessels on the high seas. Thus, there is no obstacle to fishing entities acting as the equivalent of flag states in high seas enforcement. Fishing entities' role in non-flag state enforcement in high seas fisheries is clarified in the next chapter.

CHAPTER FIVE

FISHING ENTITIES IN NON-FLAG STATE ENFORCEMENT

1. Introduction

The high seas are not under the jurisdiction of any state; therefore, as Chapter 4 explains, the purpose of flag states' enforcement on the high seas is to control their vessels in order to ensure the freedom of the high seas. The objective of non-flag state enforcement, though, exceeds flag state's enforcement in order to ensure that each state can enjoy the benefits from the freedom of the high seas. Consequently, non-flag state enforcement amounts to interference with the exclusive jurisdiction of flag state; on the other, it can be regarded as a supplement to flag state's jurisdiction.[1]

In customary international law and Article 110 of UNCLOS, non-flag state enforcement on the high seas is usually limited to the circumstances of piracy, slave trading, unauthorised broadcasting, drug trafficking, ships of uncertain nationality and stateless ships.[2] With the rapid development of fishing technology, non-flag state enforcement on the high seas has extended to the field of fishing. As some flag states might be unwilling or unable to ensure compliance by their vessels with their international obligations concerning fisheries conservation and management on the high seas, non-flag state enforcement becomes an effective way to deal with this problem.[3] As the subject of non-flag state enforcement is regarded as a

---

[1] Rosemary Rayfuse even asserts that "non-flag state enforcement already forms a significant part of the jurisdiction balances in the UNCLOS." *See* Rosemary Rayfuse, "Regulation and Enforcement in the Law of the Sea: Emerging Assertions of a Right to Non-flag State Enforcement in the High Seas Fisheries and Disarmament Contexts," *Australian Year Book of International Law*, Vol. 24(August 2005), p. 183.

[2] R.R. Churchill and A.V. Lowe, *The Law of the Sea*, 3rd ed. (Manchester: Manchester University Press, 1999), pp. 209-214.

[3] Rayfuse, *supra note* 1, p. 184.

state to an extent, whether fishing entities can conduct enforcement on the high seas like a non-flag state is worth clarifying. The purpose of this chapter is to discuss the role of fishing entities in non-flag state enforcement on the high seas for fisheries conservation and management. This chapter first reviews the mode of enforcement on the high seas and the development of non-flag state enforcement in high seas fisheries. Next, this chapter determines the legal status of fishing entities in the regime of non-flag state enforcement under the UNCLOS, 1995 Fish Stock Agreement and IUU International Plan of Action. Finally, it discusses the position of fishing entities' non-flag state enforcement under bilateral and regional treaties.

2. The Mode of Enforcement on the High Seas

2.1 The Right of Approach

As a supplement to the exclusive jurisdiction of flag states, non-flag state enforcement is permitted in exceptions such as piracy and slave trading. Under the customary international law of the sea, the right of approach is usually the first step of non-flag state enforcement. It allows warships to approach a vessel on the high seas to verify its identity and nationality in order to thwart illegal activities and maintain order on the high seas.[4] A warship usually requests that a vessel show its flag.[5] If the vessel refuses to hoist her flag, then the warship can fire a blank shot.[6] Then the warship can shoot across her bow if the vessel ignores the warning.[7] If the vessel still refuses to show her flag, the warship may resort to force in order to verify her nationality.[8]

---

[4] Robert C.F. Reuland, "Interference with Non-National Ships on the High Seas: Peacetime Exceptions to the Exclusivity of Flag State Jurisdiction," *Vanderbilt Journal of Transnational Law*, Vol. 22(1989), pp. 1169–1170; also *see* Ian Brownlie, *Principles of Public International Law*, 7th ed. (New York: Oxford University Press, 2008), p. 232. At first, this right under customary international law was used to prevent and deter piracy; with the development of international law of the sea, it has gradually been expanded to other illegal activities on the high seas. *See* L. Oppenheim, *Oppenheim's International Law*, 9th ed. (eds. Robert Jennings and Arthur Watts London: Longman, 1992, Vol. I), pp. 736–739.

[5] Reuland, *supra note* 4, p. 1170.

[6] Oppenheim, *supra note* 4, p, 738; also *see* C. J. Colombos, *The International Law of the Sea*, 6th ed. (London: Longmans Green & Co. Ltd., 1967), p. 312.

[7] *Ibid.*

[8] *Ibid.*

Although in 1859 correspondence the British and American governments agreed upon the propriety of every merchant vessel showing its flag on the ocean whenever requested by the warship of any nation[9], no related regulation indicates that vessels have the obligation to show their flags to warships. However, if a vessel refuses to exhibit its flag, it creates reasonable grounds for suspecting that the vessel is engaged in illegal activities, so the warship may request to board and inspect the suspected vessel.

2.2 The Right of Boarding and Inspection

Boarding and inspection, which also falls under customary law, is the main method of high seas enforcement. It contains two elements. First, in boarding, a warship sends a boat to the suspected vessel and checks her documents to verify the legitimacy of the vessel and her flag. Second, in inspection, if a warship after boarding has reasonable grounds to suspect that the suspected vessel is not entitled to the protection of the flag which she has hoisted or that the suspected vessel might have engaged in proscribed activities, then the warship may proceed to a further examination of the vessel. The purpose of inspection is to find the evidence that supports the suspicion of the warship.[10] The master and crew of the inspected vessel do not have obligation to render any assistance, except for example to open locked cupboards.[11]

This customary right is codified in the 1958 Convention on the High Seas. Article 22(2) of the Convention on the High Seas states that, "[i]n the cases [in which a vessel is reasonably suspected of being engaged in some proscribed activities], the warship may proceed to verify the ship's right to fly its flag. To this end, it may send a boat under the command of an officer to the suspected ship. If suspicion remains after the documents have been checked, it may proceed to a further examination on board the ship." This right is also codified in the UNCLOS under the exact same clause as Article 110(2). According to customary law, if the suspected vessel is proven innocent, the inspectors must replace anything removed and make a memorandum in the log-book, then, the inspected vessel may continue on the high seas as planned.[12] In addition, the flag state of the

---

[9] *Ibid.*
[10] Reuland, *supra note* 4, pp. 1175–1176.
[11] Oppenheim, *supra note* 4, p. 738.
[12] *Ibid.*

warship should compensate any loss or damage sustained by the vessel as a result of the boarding and inspection.[13]

2.3 Arrest

Following boarding and inspection, if the suspected vessel has committed or is found to have engaged in proscribed activities based on sufficient evidence, then the warship may arrest the vessel.[14] During the arrest, the warship should keep the vessel and its cargo safe and intact.[15] There is no clear regulation of jurisdiction over the arrested vessel. Oppenheim writes:

> The arrested vessel, either accompanied by the arresting vessel or not, must be brought to such harbour as is determined by the cause of the arrest.[16]

The provisions in the 1958 Convention on the High Seas and UNCLOS concerning enforcement on high seas go no further than boarding and inspection. Neither has regulations regarding the right of arrest or provides the next step for warships and inspected vessel when the vessel is found to have engaged in proscribed activities.

3. The Expansion of Enforcement on the High Seas

The contents of enforcement on the high seas have become broader than in the past. However, the 1958 Convention on the High Seas and UNCLOS do not have provisions concerning the right of approach; they only regulate the rights of boarding and inspection. The concept of boarding and inspection stems from a convention prohibiting the slave trade between Britain and France in 1845.[17] The provisions concerning transportation of liquor in the North Sea in the 1887 Hague Convention[18] endowed the warships of contracting parties with the right to stop a vessel hoisting the flag of contracting parties and the right to board and inspect

---

[13] *See* Article 22(3) of the 1958 Convention on the High Seas and Article 110(3) of UNCLOS.

[14] Reuland, *supra note* 4, pp. 1176.

[15] Oppenheim, *supra note* 4, p, 739.

[16] *Ibid.*

[17] René-Jean Dupuy and Daniel Vignes, eds., *A Handbook on the New Law of the Sea* (Boston: Martinus Nijhoff, 1991), p. 420.

[18] The full name of the convention is the Convention Respecting the Liquor Traffic in the North Sea. The full text is available at *http://www.legislation.gov.uk/ukpga/Vict/56-57/17/schedule* (visited on 5/27/2013).

the documents of that vessel.[19] The system of boarding and inspection on the high seas by non-flag states breaches the exclusive jurisdiction of flag states in customary international law; therefore, it was initially limited under specific agreements between states.

At turn of the nineteenth century, Britain tried to board and inspect an American vessel in order to recruit navy crewmembers, but United States objected, leading to a conflict[20] which has spurred many discussions. Pitman B. Potter thought that a state had no right to command another state's vessel to stop or to board and inspect this vessel during peace-time unless that vessel was engaged in piracy or slave trade.[21] Professor H.A. Smith considered that under customary law, except the right of approach, a state could not interfere with another state's vessel on the high seas during peace-time.[22] He believed that, unless the vessel was engaged in piracy or other improper activities, any interference by a non-flag state must be permitted by a treaty.[23] Charles Cheney Hyde took a somewhat different point of view. He contended that the right of boarding and inspection on the high seas by non-flag states could be used only during war-time; no state possessed this right during peace-time.[24] Gilbert Gidel had a similar position as Hyde; he thought that, during peace-time, a state could not interfere with another state's vessel on the high seas according to the

[19] Dupuy and Vignes, *supra note 17.*

[20] John Bassett Moore, *A Digest of International Law* (Ann Arbor: UMI, A Bell & Howell Company, 1993, Vol. 2), pp. 987–1001.

[21] *See* Myres S. McDougal and William T. Burke, *The Public Order of the Oceans: A Contemporary International Law of the Sea* (New Haven: Yale University Press, 1962), p. 887; Pitman B. Potter, *The Freedom of the Seas, in History, Law and Politics* (London: Longmans, Green & Co., 1924), p. 101.

[22] H.A. Smith, *The Law and Custom of the Sea*, 3rd ed. (London: Stevens and Sons Limited, 1959), pp. 64-65.

[23] *Ibid.* Smith states: "This 'right of approach' (*verification du pavillon or reconnaissance*) is the only qualification under customary law of the general principle which forbids any interference in time of peace with ships of another nationality upon the high seas. Any other act of interference (apart from the repression of piracy) must be justified under powers conferred by treaty. Provided that the merchant vessel responds by showering her flag the captain of the warship is not justified in boarding her or taking any further action, unless there is reasonable ground for suspecting that she is engaged in piracy or some other improper activity."

[24] Charles Cheney Hyde, *International Law: Chiefly as Interpreted and Applied by the United States* 2nd rev. ed. (Boston: Little, Brown and Company, 1951, Vol. 1), p. 764.

principles of freedom of the high seas.[25] Furthermore, in an International Law Commission Report, J.P.A. François asserted that international law holds that, during peace-time, the only control measure that non-flag states may practice is the right of approach, which means the right to ascertain the identity and nationality of the vessels but not the right to inspect the documents of the vessels or the right of search.[26] At the seventh meeting of the International Law Commission in 1955, the final provisions concerning boarding and inspection in the 1958 Convention on the High Seas were drafted.[27]

However, from the customary international law and the 1958 Convention on the High Seas to UNCLOS, the contents of enforcement on high seas has become broader. The subject of enforcement in customary law and the 1958 Convention is the "warship;" but in UNCLOS, it is expanded to "any other duly authorized ships clearly marked and identifiable as being on government service." Additionally, the object of enforcement was also enlarged from "a foreign merchant ship" in the 1958 Convention to "a foreign ship" in UNCLOS.[28] In addition, according to

---

[25] McDougal and Burke, *supra note* 21, p. 888. For the full text, *see*: Gilbert Gidel, "Memorandum," in United Nations Secretariat, *Memorandum on the Regime of the High Seas* (U.N. Doc. No. A/CN.4/32) (1950), available at http://untreaty.un.org/ilc/documentation/english/a_cn4_32.pdf (visited on 10/11/2011).

[26] *Ibid.*, p. 889. For the full text, *see*: J.P.A. François, *Report on the Regime of the High Seas* (U.N. Doc. No. A/CN.4/17, 1950), available at http://untreaty.un.org/ilc/documentation/english/a_cn4_17.pdf (visited on 13/12/2011). François reiterated his position in the *Second Report on the Regime of the High Seas* (U.N. Doc. No. A/CN.4/42, 1951), *available at http://untreaty.un.org/ilc/documentation/english/a_cn4_42.pdf* (visited on 11/11/2011). Serving on the International Law Commission in 1951, Judge Manley O. Hudson suggested that the rights of boarding and inspecting vessels suspected of engaging in piracy and vessels suspected of engaging in slave trade should not be distinguished but combined. *See* International Law Commission, *Yearbook of the International Law Commission* (New York: United Nations, 1951, Vol. 1), p. 354.

[27] McDougal and Burke, *supra note* 21, p. 890.

[28] Under the 1958 convention, any ships which are not the merchant ships are excluded from enforcement. According to this stipulation, ships which belong to a government with commercial purpose can also be the object of enforcement; *see* McDougal and Burke, *supra note* 21, p. 892. During the drafting of the 1958 Convention, applying enforcement by warships to ships suspected of engaging in hostile acts causing imminent danger to the warship state was considered. This provision was not adopted was not primarily because the warship had no right to conduct this enforcement but, rather, to prevent states from abusing this provision as the concepts of "imminent danger" and "hostile acts" are too ambiguous to

the 1958 Convention, the ships may become the object of enforcement on the high seas because they are suspected of engaging in piracy, slave trade, flying a foreign flag or refusing to show their flag but are actually of the same nationality as the warship; however, in UNCLOS, the warships may also conduct enforcement against ships which are without nationality or engaged in unauthorised broadcasting.

In addition to these above activities which may justify non-flag state enforcement, enforcement on the high seas is expanded to fishing activities.[29] Due to the necessity for conservation and management of high seas fisheries, many international agreements and RFMOs contain related measures. One method to ensure the implementation of these conservation measures is enforcement on the high seas. In traditional international law of the sea, flag state enforcement, even on the high seas, is the norm. However, due to the urgency of declining fishery resources, non-flag state enforcement on the high seas to ensure compliance with the conservation and management measures is developing. However, it is as of yet only a treaty right. It is regulated in the UNCLOS and specified in the UNFSA, as well as by some RFMOs. A discussion of non-flag state enforcement on the high seas for fisheries follows.

4. The Legal Status of Non-Flag State Enforcement in High Seas Fisheries

4.1 Global Instruments

A. United Nations Convention of the Law of the Sea
Although the UNCLOS does not directly permit enforcement to be applied to ships engaged in proscribed fishing activities, Article 116 preserves the

---

clarify and interpret. *See* E.D. Brown, *The International Law of the Sea* (Aldershot: Dartmouth Publishing Company, 1994, Vol. 1), p. 314.

[29] In addition to fisheries, another activity which necessitates enforcement by non-flag states on the high seas is anti-proliferation of the weapons of mass destruction (WMDs). After the 9/11 attacks, the United States proposed the concept of proliferation security initiative (PSI) in 2002. This concept permits states to take collective cooperative action to prevent the proliferation of WMDs. One method they may use is to board, inspect and even seize vessels suspected of transporting WMDs on the high seas. Regarding PSI, please *see* Michael Byers, "Policing the High Sea: The Proliferation Security Initiative," *American Journal of International Law*, Vol. 98, No. 3(July, 2004), pp. 526-545; "Asia: Practising to provoke: Counter-proliferation," *The Economist*, Vol. 368(Sept. 20, 2003), p. 79; Rayfuse, *supra note* 1, pp. 181–200.

possibility that the right of non-flag state enforcement concerning fishing may be endowed by treaties. Article 117 further indicates that states should take necessary measures to conserve and manage the living resources of the high seas. It does not provide any concrete measures for states but can be regarded as implying that such measures could include enforcement measures.

The UNCLOS does not directly address non-flag state's enforcement for fishing on the high seas but maintains the possibility. Therefore, non-flag state's boarding and inspection procedure for fishery conservation and management on the high seas requires separate treaties. In this position, the role of non-flag state's enforcement on the high seas for fishing entities is not be influenced by its legal status in the UNCLOS. In other words, as UNCLOS does not inhibit the possibility of non-flag state enforcement on the high seas for fisheries by treaty, fishing entities may, like states, possess the right to participate in this enforcement regime through treaties or RFMOs.

B. The United Nations Fish Stocks Agreement

The UNFSA regulates the conservation and management of straddling and highly migratory fish stocks on the high seas. To achieve these aims, it emphasises the obligation of contracting parties to cooperate and help each other in enforcement measures. Articles 20, 21 and 22 of the UNFSA are the core provisions regarding high seas enforcement. Article 20 indicates that states should cooperate in enforcement at the international level. Article 20(6) allows coastal states which are authorised by flag states to conduct enforcement against suspected unauthorised fishing vessels: "Where there are reasonable grounds for believing that a vessel on the high seas has been engaged in unauthorized fishing within an area under the jurisdiction of a coastal State, the flag State of that vessel, at the request of the coastal State concerned, shall immediately and fully investigate the matter. The flag State shall cooperate with the coastal State in taking appropriate enforcement action in such cases and may authorize the relevant authorities of the coastal State to board and inspect the vessel on the high seas."

Article 21 regulates states' cooperation in enforcement at the subregional and regional levels. Articles 21(1) and (2) allows states in RFMOs to board and inspect each other's vessels according to the enforcement procedures adopted by RFMOs to ensure compliance with the conservation and management measures for straddling and highly migratory fish stocks.[30] Under Article 21(4), inspecting states should inform all states whose vessels fish on the high seas of the form of identification issued to their duly authorised inspectors, and the vessels used for boarding and inspection should be clearly marked and identifiable as in government service.

The mode of enforcement described in the UNFSA is boarding and inspection. Any further actions, such as arrest, after finding evidence of violations by non-flag states are still taken over by flag states. Pursuant to Article 21(5) of the UNFSA, if there are clear grounds for believing that a vessel has engaged in any activity contrary to conservation and management measures, the inspecting state should promptly notify and deliver evidence of the violation to the flag state. The flag state should respond to the notification within three working days by either conducting an investigation and sending the findings to the inspecting state or by authorising the inspecting state to investigate.[31] If the flag state chooses to authorise the inspecting state to investigate the alleged violation, the authorised state should return the results of investigation to the flag state without delay; the flag state is obliged to undertake further enforcement involving the vessel. Otherwise, the flag state may authorise the inspecting state to take such enforcement action as the flag state may specify with respect to the vessel.[32] However, if after boarding and inspecting there are clear grounds for believing that a vessel has committed a "serious

---

[30] Article 21(1) states: "In any high seas area covered by a subregional or regional fisheries management organization or arrangement, a State Party which is a member of such organization or a participant in such arrangement may, through its duly authorized inspectors, board and inspect, in accordance with paragraph 2, fishing vessels flying the flag of another State Party to this Agreement, whether or not such State Party is also a member of the organization or a participant in the arrangement, for the purpose of ensuring compliance with conservation and management measures for straddling fish stocks and highly migratory fish stocks established by that organization or arrangement." Article 21(2) reads: "States shall establish, through subregional or regional fisheries management organizations or arrangements, procedures for boarding and inspection pursuant to paragraph 1, as well as procedures to implement other provisions of this article. Such procedures shall be consistent with this article and the basic procedures set out in article 22 and shall not discriminate against non-members of the organization or non-participants in the arrangement. Boarding and inspection as well as any subsequent enforcement action shall be conducted in accordance with such procedures. States shall give due publicity to procedures established pursuant to this paragraph."

[31] *See* Article 21(6) of the UNFSA.

[32] *See* Article 21(7) of the UNFSA.

violation"[33] and the flag state does not respond or take further action, then Article 21(8) permits inspectors to "remain on board and secure evidence" and "require the master to assist in further investigation including, where appropriate, by bringing the vessel without delay to the nearest appropriate port." Although the flag state could choose to authorise the inspecting state to investigate, it can also withdraw this authorisation at any time. Article 21(12) states that, "[n]otwithstanding the other provisions of this article, the flag State may, at any time, take action to fulfil its obligations … with respect to an alleged violation. Where the vessel is under the direction of the inspecting State, the inspecting State shall, at the request of the flag State, release the vessel to the flag State along with full information on the progress and outcome of its investigation." However, Article 21(17) allows any states not only to board and inspect a stateless vessel but also to conduct follow-up enforcement if they find evidence of violations.[34]

The UNFSA encourages RFMOs to establish their own boarding and inspection procedures. If they have not yet, Article 22 of the UNFSA

---

[33] Article 21(11) further defines a serious violation as:

"(a) fishing without a valid licence, authorization or permit issued by the flag State in accordance with article 18, paragraph 3 (a);

(b) failing to maintain accurate records of catch and catch-related data, as required by the relevant subregional or regional fisheries management organization or arrangement, or serious misreporting of catch, contrary to the catch reporting requirements of such organization or arrangement;

(c) fishing in a closed area, fishing during a closed season or fishing without, or after attainment of, a quota established by the relevant subregional or regional fisheries management organization or arrangement;

(d) directed fishing for a stock which is subject to a moratorium or for which fishing is prohibited;

(e) using prohibited fishing gear;

(f) falsifying or concealing the markings, identity or registration of a fishing vessel;

(g) concealing, tampering with or disposing of evidence relating to an investigation;

(h) multiple violations which together constitute a serious disregard of conservation and management measures; or

(i) such other violations as may be specified in procedures established by the relevant subregional or regional fisheries management organization or arrangement."

[34] Article 21(17) reads: "Where there are reasonable grounds for suspecting that a fishing vessel on the high seas is without nationality, a State may board and inspect the vessel. Where evidence so warrants, the State may take such action as may be appropriate in accordance with international law."

details basic boarding and inspection procedures which can be adopted directly by RFMOs or as transitional rules. It specifies that authorised inspectors should "present credentials to the master of the vessel and produce a copy of the text of the relevant conservation and management measures; ... initiate notice to the flag State at the time of the boarding and inspection; not interfere with the master's ability to communicate with the authorities of the flag State during the boarding and inspection; provide a copy of a report on the boarding and inspection to the master and to the authorities of the flag State, noting therein any objection or statement which the master wishes to have included in the report; promptly leave the vessel following completion of the inspection if they find no evidence of a serious violation; and avoid the use of force. .... [If necessary,] the degree of force used shall not exceed that reasonably required in the circumstances." Pursuant to Article 22(2), the inspectors have the authority to inspect the vessel and its licence, gear, equipment, records, facilities, fish and fish products and any relevant documents necessary to verify compliance with the relevant conservation and management measures. The master of the vessel which is requested to be inspected should fully cooperate with the inspectors.[35] If the master of the vessel refuses to be inspected, then the flag state should direct the master to immediately accept the boarding and inspection request. If the master of the vessel still refuses, the flag state should suspend the vessel's authorisation to fish and order the vessel to return immediately to port.[36]

Although the state is the subject of Articles 20-22 in the UNFSA regarding non-flag state enforcement in high seas fisheries, fishing entities are not excluded from its provisions. Article 1(3) states that it applies *mutatis mutandis* to other fishing entities whose vessels fish on the high seas, and pursuant to Article 17(3), these fishing entities should be

---

[35] Article 22(3) states: "The flag State shall ensure that vessel masters:

(a) accept and facilitate prompt and safe boarding by the inspectors;
(b) cooperate with and assist in the inspection of the vessel conducted pursuant to these procedures;
(c) do not obstruct, intimidate or interfere with the inspectors in the performance of their duties;
(d) allow the inspectors to communicate with the authorities of the flag State and the inspecting State during the boarding and inspection;
(e) provide reasonable facilities, including, where appropriate, food and accommodation, to the inspectors; and
(f) facilitate safe disembarkation by the inspectors."

[36] The flag state shall advise the inspecting state of the action it has taken when the circumstances arise. *See* Article 22(4) of the UNFSA.

requested to cooperate with RFMOs in implementing conservation and management measures. Although the UNFSA allows no door for fishing entities to sign or accede to it, it encourages fishing entities to be brought into RFMOs so as to apply conservation and management measures as extensively as possible to fishing activities in the relevant area.[37] The UNFSA stipulates that each RFMO should establish boarding and inspection procedures which can be applied to members in the relevant high seas areas. It also encourages the fishing entities whose vessels fish on the high seas to join or become members of RFMOs. From this position, the UNFSA does not exclude the possibility of fishing entities' participating in non-flag state enforcement on the high seas. In other words, the UNFSA is not violated if a RFMO adopts boarding and inspection procedures allowing all of its members, including fishing entities, to implement its high seas enforcement provisions.

C. Illegal, Unreported and Unregulated Fishing International Plan of Action[38]

The provisions concerning non-flag state enforcement on the high seas in the IUU International Plan of Action are not as specific as those in the UNFSA but only respond to the UNFSA's appeal to carry out the related boarding and inspection regimes. According to Paragraph 24.10 of the IUU International Action Plan, states "should undertake comprehensive and effective monitoring, control and surveillance (MCS) of fishing from its commencement … by ensuring effective implementation of national and … internationally agreed boarding and inspection regimes consistent with international law, recognizing the rights and obligations of masters and of inspection officers, and noting that such regimes are provided for in certain international agreements, such as the 1995 [UNFSA]." In addition to the UNFSA, paragraph 80.8 reiterates the need to establish boarding and inspection procedures within RFMOs: "States, acting through relevant regional fisheries management organizations, should take action to strengthen and develop innovative ways … to prevent, deter, and eliminate IUU fishing. Consideration should be given to … development within a regional fisheries management organization … of boarding and inspection regimes consistent with international law, recognizing the rights and obligations of masters and inspection officers."

---

[37] Fishing entities' legal status under the UNFSA is discussed in Chapter 2 of this book.

[38] Also *see* Chapter 1 of this book.

As the IUU International Action Plan is a voluntary instrument that applies to all states and entities and to all fishers, it poses not obvious challenge to the legal status of fishing entities. Furthermore, the regulations concerning enforcement on the high seas in the IUU International Action Plan mainly reiterates and consents to the boarding and inspection regimes in the UNFSA and the establishment of similar regimes by RFMOs. This plan does not affect fishing entities' status in RFMOs or apply to them the UNFSA's provisions for non-flag state enforcement in high seas fisheries.

4.2 Regional and Bilateral Treaties

A. Regional Treaties

In 1882, the International Convention for Regulating the Police of the North Sea Fisheries[39] (the 1882 North Sea Fisheries Convention) was signed to regulate the policing of the fisheries in the North Sea outside territorial waters.[40] At that time, enforcement on the high seas was still largely limited to flag states. Pursuant to Article XXIX, the commanders of cruisers could only require a vessel suspected of violating the provisions of the convention to show her official document establishing her nationality.[41] The commanders of cruisers could board and inspect a vessel which was not of their own nationality only provided that it was "necessary for the purpose of obtaining proof of an offence or of a contravention of regulations respecting the police of the fisheries."

In the twentieth century, non-flag state enforcement on the high seas is not as limited as in the nineteenth century. In 1952, the United States, Canada and Japan concluded the International Convention for the High

[39] The parties were: Belgium, Denmark, France, Germany, Great Britain and Netherlands. The full text is available at *http://iea.uoregon.edu/pages/view_treaty.php?t=1882-PoliceNorthSeasFishery.EN.txt&par=view_treaty_html* (visited on 13/12/2011).

[40] At that time, the concept of the exclusive economic zone did not exist. The high seas meant all waters outside the 3-mile limit from low-water mark along the entire coast of a country.

[41] Article XXIX of the 1882 North Sea Fisheries Convention reads: "When the commanders of cruizers have reason to believe that an infraction of the provisions of the present Convention has been committed, they may require the master of the boat inculpated to exhibit the official document establishing her nationality. The fact of such document having been exhibited shall then be endorsed upon it immediately."

Seas Fisheries of the North Pacific Ocean[42], in which Article X allows the contracting parties to board each other's vessels to inspect equipment, books and documents and question those on board.[43] If there are reasonable grounds to believe that the inspected vessel was engaged in operation in violation of this convention, then the inspecting official may arrest or seize such person or vessel.[44] However, the arrested or seized person or vessel should be handed over to the flag state; *i.e.* only the flag state had the right of jurisdiction.[45] On 25 April 1978, the Protocol

---

[42] The Convention entered into force on 12 June 1953. The full text of the convention can be found on the website of the Center for International Earth Science Information Network (CIESIN):
http://sedac.ciesin.columbia.edu/entri/texts/fisheries.north.pacific.1952.html (visited on 19/12/2012). In accordance with this convention, the International North Pacific Fisheries Commission (INPFC) was established in 1952 with Canada, Japan and the United States of America as members, to contribute to understanding of the life history and distribution of anadromous species, groundfish, crab and marine mammals in the North Pacific Ocean and Bering Sea. The INPFC was dissolved when the Convention for the Conservation of Anadromous Stocks in the North Pacific Ocean entered into force on 16 February 1993. Regarding the INPFC, please *see http://www.npafc.org/new/ipnfc.html* (visited on 19/12/2012).

[43] Article 10(1)(a) of the International Convention for the High Seas Fisheries of the North Pacific Ocean reads: "When a fishing vessel of a Contracting Party has been found in waters in which that Party has agreed to abstain from exploitation in accordance with the provisions of this Convention, the duly authorized officials of any Contracting Party may board such vessel to inspect its equipment, books, documents, and other articles and question the persons on board. Such officials shall present credentials issued by their respective Governments if requested by the master of the vessel."

[44] Article 10(1)(b) of the International Convention for the High Seas Fisheries of the North Pacific Ocean further states: "When any such person or fishing vessel is actually engaged in operation in violation of the provisions of this Convention, or there is reasonable ground to believe was obviously so engaged immediately prior to boarding of such vessel by any such official, the latter may arrest or seize such person or vessel. In that case, the Contracting Party to which the official belongs shall notify the Contracting Party to which such person or vessel belongs of such arrest or seizure and shall deliver such vessel or persons as promptly as practicable to the authorized officials of the Contracting Party to which such vessel or person belongs at a place to be agreed upon by both Parties. Provided, however, that when the Contracting Party which receives such notifications cannot immediately accept delivery and makes request, the Contracting Party which gives such notification may keep such person or vessel under surveillance within its own territory under the conditions agreed upon by both of the Contracting Parties."

[45] Article 10(1)(c) of the International Convention for the High Seas Fisheries of the North Pacific Ocean reads: "Only the authorities of the Party to which the

Amending the International Convention for the High Seas Fisheries of the North Pacific Ocean[46] (the 1978 Amended Protocol) was signed in Tokyo. The provisions concerning non-flag state enforcement on the high seas in the 1978 Amended Protocol were similar to the 1952 Convention but specified that the inspected vessels were "vessels fishing for anadromous species." As well, the Amended Protocol permits the inspecting official to not only arrest and seize the suspected person or vessel but also to investigate further, the circumstances warrant it.[47] The Amended Annex of the 1978 Amended Protocol contained a memorandum of understanding on Enforcement, which recorded details of enforcement between Japan and United States in the Japanese land-based fishery area, including the numbers of patrol vessels, precise area of enforcement, and other acts of cooperation.[48] In this memorandum, Japan and the United States agree to accommodate an observer of the other government aboard its patrol vessels assigned to enforce along the eastern limit of the Japanese land-based fishery area for up to four weeks.[49] Such observers would be paid by their government and could communicate with their parent agency with the consent of the captain of the patrol vessel.[50] Furthermore, the observer should comply with instructions of the host enforcement officials under all circumstances.[51] Most importantly, the observer would not exercise any enforcement authority but only observe the activities of the enforcement officials of the patrol vessel.[52]

As well, in 1967, the Convention on Conduct of Fishing Operations in the North Atlantic[53] was adopted to ensure order in the fishing grounds of

above-mentioned person or fishing vessel belongs may try the offence and impose penalties therefor. The witnesses and evidence necessary for establishing the offence, so far as they are under the control of any of the Parties to this Convention, shall be furnished as promptly as possible to the Contracting Party having jurisdiction to try the offence."

[46] The amended protocol entered into force on 15 February 1979. The full text of the protocol can be found at *http://www.npafc.org/new/inpfc/INPFC%20_convention.pdf* (visited on 03/01/2012).

[47] *See* Article XI of the 1978 amended protocol.

[48] *See* the amended protocol, *supra note* 46.

[49] *Ibid.*

[50] *Ibid.*

[51] *Ibid.*

[52] *Ibid.*

[53] The parties to the convention are the governments of Belgium, Canada, Denmark, the French Republic, the Federal Republic of Germany, Iceland, Ireland, Italy, Luxembourg, the Netherlands, Norway, the Polish People's Republic, Portugal, Spain, Sweden, the Union of Soviet Socialist Republics, the United

the North Atlantic area. Article 9 stipulates that, if an authorised officer has reason to believe that a vessel of any contracting party is not complying with the provisions of the convention, he may order the vessel to stop and board it for enquiry and report. However, this non-flag state enforcement is permitted only if no authorised official of the flag state available can do so.[54]

The International Commission for the Northwest Atlantic Fisheries (ICNAF)[55] in 1970 adopted a Scheme of Joint International Enforcement of the fishery regulations in the Convention Area[56], which allows the contracting parties to appoint inspectors to conduct boarding and inspection of members' vessels in the convention area. Subsequently, Article XVIII of the Convention on Future Multilateral Cooperation in the Northwest Atlantic Fisheries[57], which was adopted on 24 October 1978 in Ottawa, also announces plans to establish a scheme including "provision for reciprocal rights of boarding and inspection by the Contracting Parties and for flag State prosecution and sanctions on the basis of evidence resulting from such boardings and inspections." Next, a Proposal for Amendment of the Scheme of Joint International Enforcement of the

---

Kingdom of Great Britain and Northern Ireland, and the United States of America. The convention entered into force on 26 September 1976. The full text can be seen at *http://www.fco.gov.uk/resources/en/pdf/3706546/3892723/21060066/Tr-Fish NorthAtlan-Ts40.1977* (visited on 03/01/2012). Regarding this convention, also *see* D. W. Van Lynden, "The Convention on Conduct of Fishing Operations in the North Atlantic," *Netherlands International Law Review*, Vol. 14(1967), pp. 245-258.

[54] *Ibid.* Article 9(12) reads: "An authorised officer shall not exercise his powers to board a vessel of another Contracting Party if an authorised officer of that Contracting Party is available and in a position to do so himself."

[55] The ICNAF was formed in 1949 to conserve and manage fishery resources in Northwest Atlantic. It was ended and replaced by the Northwest Atlantic Fisheries Organization (NAFO) in 1979. *See http://www.nafo.int/icnaf/frames/icnaf.html* (visited on 04/12/2012).

[56] The scheme came into effect on 7 January 1971, and its operation by all contracting parties commenced on 1 July 1971. The full text of the scheme can be found in the *ICNAF Annual Proceedings*, Vol. 20(1969-1970), pp. 20–22. Also *see* Rosemary Gail Rayfuse, *Non-Flag State Enforcement in High Seas Fisheries* (Leiden: Martinus Nijhoff Publishers, 2004), pp. 234–236.

[57] The Convention came into force on 1 January 1979, establishing the NAFO to replace the ICNAF. The prime objective of NAFO is to contribute through consultation and cooperation to the optimum utilisation, rational management and conservation of the fishery resources of the convention area. The full text of the convention can be found on the official NAFO website,
http://www.nafo.int/about/frames/about.html (visited on 05/01/2012).

Fishery Regulations in Regulatory Area[58] was adopted on 7 June 1979 to succeed the ICNAF International Enforcement Scheme.

The NPAFC replaced the International North Pacific Fisheries Commission (INPFC), established by the 1952 International Convention for the High Seas Fisheries of the North Pacific Ocean. The NPAFC's provisions concerning non-flag state enforcement followed those in the 1978 Amended Protocol, including policies for boarding, inspection, arrest, seizure, investigation and judicial jurisdiction over the inspected vessel.[59]

---

[58] The full text of the proposal can be found in the *NAFO Annual Report*, Vol. 1(1979), pp. 70–71.

[59] Article IX of the Convention of NPAFC states:

" ….(i) The duly authorized officials of any Contracting Party may board vessels fishing for anadromous species of the other Contracting Parties to inspect equipment, logs, documents, catch and other articles and question the persons on board for the purpose of carrying out the provisions of this Convention. Such inspections and questioning shall be made so that the vessels suffer the minimum interference and inconvenience. Such officials shall present credentials issued by their respective Governments if requested by the master of the vessel.

(ii) When any such person or fishing vessel is actually engaged in operations in violation of the provisions of this Convention, or there is reasonable ground to believe was obviously so engaged prior to boarding of such vessel by any such official, the latter may arrest or seize such person or vessel and further investigate the circumstances if necessary. The Contracting Party to which the official belongs shall notify promptly the Contracting Party to which such person or vessel belongs of such arrest or seizure, and shall deliver such person or vessel as promptly as practicable to the authorized officials of the Contracting Party to which such person or vessel belongs at a place to be agreed upon by both Parties. Provided, however, that when the Contracting Party which receives such notification cannot immediately accept delivery, the Contracting Party which gives such notification may keep such person or vessel under surveillance within the waters of the Convention area or within its own territory under the conditions agreed upon by both the Contracting Parties.

(iii) Only the authorities of the Contracting Party to which the above-mentioned person or fishing vessel belongs may try the offense and impose penalties therefor. The witnesses and evidence necessary for establishing the offense, so far as they are under the control of any of the Contracting Parties to this Convention, shall be furnished as promptly as possible to the Contracting Party having jurisdiction to try the offense and shall be taken into account, and utilized as appropriate, by the executive authority of that Contracting Party having jurisdiction to try the offense."

In 1975, the ICCAT adopted a Scheme of Joint International Inspection[60] under which non-flag state boarding and inspection could be conducted by the contracting governments. In 2006, this scheme was reiterated through the Recommendation by ICCAT to Establish a Multi-Annual Recovery Plan for Bluefin Tuna in the Eastern Atlantic and Mediterranean.[61] This recommendation states that, in the framework of the multi-annual management plan for bluefin tuna, the contracting parties, cooperating non-contracting parties, entities or fishing entities agreed to apply this scheme.[62] However, no practice under the scheme has been reported.[63]

In the Convention on the Conservation and Management of the Pollock Resources in the Central Bering Sea[64] signed on 16 June 1994 in Washington and entering into force on 8 December 1995, Article XI(6) has provisions concerning non-flag state enforcement: "(a) Each Party consents to the boarding and inspection of fishing vessels flying its flag and located in the Convention Area by duly authorized officials of any other Party for compliance with this Convention or measures. … (b) Such officials may inspect the vessel, … catch, fishing gear, and relevant documents and logbooks, and question the master, the fishing master, and other officers on board."[65]

At present, the WCPFC offers the most complete and detailed regulations concerning non-flag state enforcement on the high seas. Article 25(6) of the WCPFC Convention states that, if there are reasonable grounds for believing that a fishing vessel on the high seas has engaged in unauthorised fishing in an area under the national jurisdiction of a WCPFC member, the flag state of the vessel should either immediately investigate the matter and cooperate with the member concerned in taking appropriate enforcement action or authorise the relevant authorities to board and inspect the vessel on the high seas. In this provision, non-flag state

---

[60] The full text of the scheme can be found in the Report of the Fourth Regular Meeting of the ICCAT, available at *http://www.iccat.int/Documents/BienRep/REP_EN_74-75_II.pdf* (visited on 17/12/2012), at Annexe 7, Appendix II, pp. 76-79.

[61] The full text of the recommendation can be found in the 15th Special Meeting of the ICCAT, available at *http://www.iccat.int/Documents/BienRep/REP_EN_06-07_I_1.pdf* (visited on 17/12/2012), pp. 130-139.

[62] *Ibid.*, p. 131.

[63] Also *see* Douglas Guilfoyle, *Shipping Interdiction and the Law of the Sea* (New York: Cambridge University Press, 2009), pp. 112-115.

[64] The full text can be seen at *http://www.afsc.noaa.gov/REFM/CBS/Docs/Convention%20on%20Conservation%20of%20Pollock%20in%20Central%20Bering%20Sea.pdf* (visited on 05/01/2012).

[65] *Ibid.*

enforcement is based on the consent and request of the flag state. Article 26 further requires the WCPFC to adopt a boarding and inspection procedure so that non-flag states can conduct enforcement without needing to obtain the consent of the flag state in advance; in other words, the procedure amounts to consent in advance.

As Article 26 of the Convention, the Western and Central Pacific Fisheries Commission Boarding and Inspection Procedures[66] (the WCPFC Boarding and Inspection Procedure) was adopted at the Third Regular Session of the WCPFC in Apia, Samoa, 11–15 December 2006. Article 17 of the procedure requires information on authorised inspectors and inspecting vessels to be open and continuingly updated.[67] Additionally, Article 13 requires that the inspecting authority provide the following documents to the WCPFC: the details of the inspecting vessel and name of the inspection authority, notification that the inspection vessel is clearly marked and identifiable as in government service and notification that the crew has received and completed training in carrying out the boarding and inspection procedures and are familiar with the fishing activities to be inspected and the provisions of the conservation and management measures.[68] Article 13 implies that the inspectors and vessels will perform enforcement only under strict conditions in order to protect the traditional right of enforcement of the flag state and prevent the abuse of non-flag state enforcement.

Pursuant to Article 20, before proceeding with boarding and inspection, the inspecting vessel should:

a. make best efforts to establish contact with the fishing vessel by radio, by the appropriate International Code of Signals or by other accepted means of alerting the vessel;
b. provide the information to identify itself as an authorised inspection vessel - name, registration number, international radio call sign and contact frequency;
c. communicate to the master of the vessel its intention to board and inspect the vessel under the authority of the Commission and pursuant to these procedures; and
d. initiate notice through the authorities of the inspection vessel of the boarding and inspection to the authorities of the fishing vessel.[69]

---

[66] The full text of the procedure can be found on the official WCPFC website, *http://www.wcpfc.int/doc/cmm-2006-08/western-and-central-pacific-fisheries-commission-boarding-and-inspection-procedures* (visited on 05/01/2012).
[67] *Ibid.*
[68] *Ibid.*
[69] *Ibid.*

Article 24 describes the obligations of inspectors while undertaking boarding and inspection:

a. present their identity card to the master of the vessel and a copy of the text of the relevant measures in force pursuant to the Convention in the relevant area of the high seas;
b. not interfere with the master's ability to communicate with the authorities of the fishing vessel;
c. complete the inspection of the vessel within 4 (four) hours unless evidence of a serious violation is found;
d. collect and clearly document any evidence they believe indicates a violation of measures in force pursuant to the Convention;
e. provide to the master prior to leaving the vessel a copy of an interim report on the boarding and inspection including any objection or statement which the master wishes to include in the report;
f. promptly leave the vessel following completion of the inspection; and
g. provide a full report on the boarding and inspection to the authorities of the fishing vessel, pursuant to paragraph 30, which shall also include any master's statement.[70]

Furthermore, Articles 23 stipulates that, in order not to constitute harassment of a fishing vessel, its officers or crew, the boarding and inspection procedure should be carried out in accordance with the internationally accepted principles of good seamanship and not interfere unduly with the lawful operation of the fishing vessel or adversely affect the quality of the catch.[71]

Article 25 further lists the obligations of the master of the inspected vessel during boarding and inspection:

a. follow internationally accepted principles of good seamanship so as to avoid risks to the safety of authorised inspection vessels and inspectors;
b. accept and facilitate prompt and safe boarding by the authorised inspectors;
c. cooperate with and assist in the inspection of the vessel pursuant to these procedures;
d. not assault, resist, intimidate, interfere with, or unduly obstruct or delay the inspectors in the performance of their duties;
e. allow the inspectors to communicate with the crew of the inspection vessel, the authorities of the inspection vessel, as well as with the authorities of the fishing vessel being inspected;
f. provide them with reasonable facilities, including, where appropriate, food and accommodation; and

[70] *Ibid.*
[71] *Ibid.*

g. facilitate safe disembarkation by the inspectors.

If the master of the fishing vessel refused to be boarded and inspected, he should offer a reasonable explanation, and the inspection authority should immediately notify the authority of the fishing vessel, as well as the commission.[72] Then, the flag state should direct the master to accept the boarding and inspection; if the master still refuses, the flag state should suspend the vessel's authorisation to fish, order the vessel to return immediately to port and immediately notify the authorities of the inspection vessel and the commission of the action it has taken in these circumstances.[73]

Articles 20–27 constitute the core of the WCPFC Boarding and Inspection Procedure and regulate the main procedures for boarding and inspection and the obligations of the inspectors, authorities of the inspection vessel, inspected vessels and the fishing vessels (see Chart 5.1).

In addition, Article 16 allows inspectors from different authorities to conduct the enforcement work on other contracting parties' inspection vessels.[74] This permission seems to break the wall separating enforcement units established by each party and to gather all inspectors and inspection vessels as a specialised department under the WCPFC.

---

[72] *See* Article 26 of the procedure.

[73] *See* Article 27 of the procedure.

[74] Article 16 of the WCPFC Boarding and Inspection Procedure reads: "To enhance the effectiveness of the Commission's boarding and inspection procedures, and to maximize the use of trained inspectors, Contracting Parties may identify opportunities to place authorized inspectors on inspection vessels of another Contracting Party.…Contracting Parties should seek to conclude bilateral arrangements to this end or otherwise facilitate communication and coordination between them for the purpose of implementing these procedures."

Chart 5.1

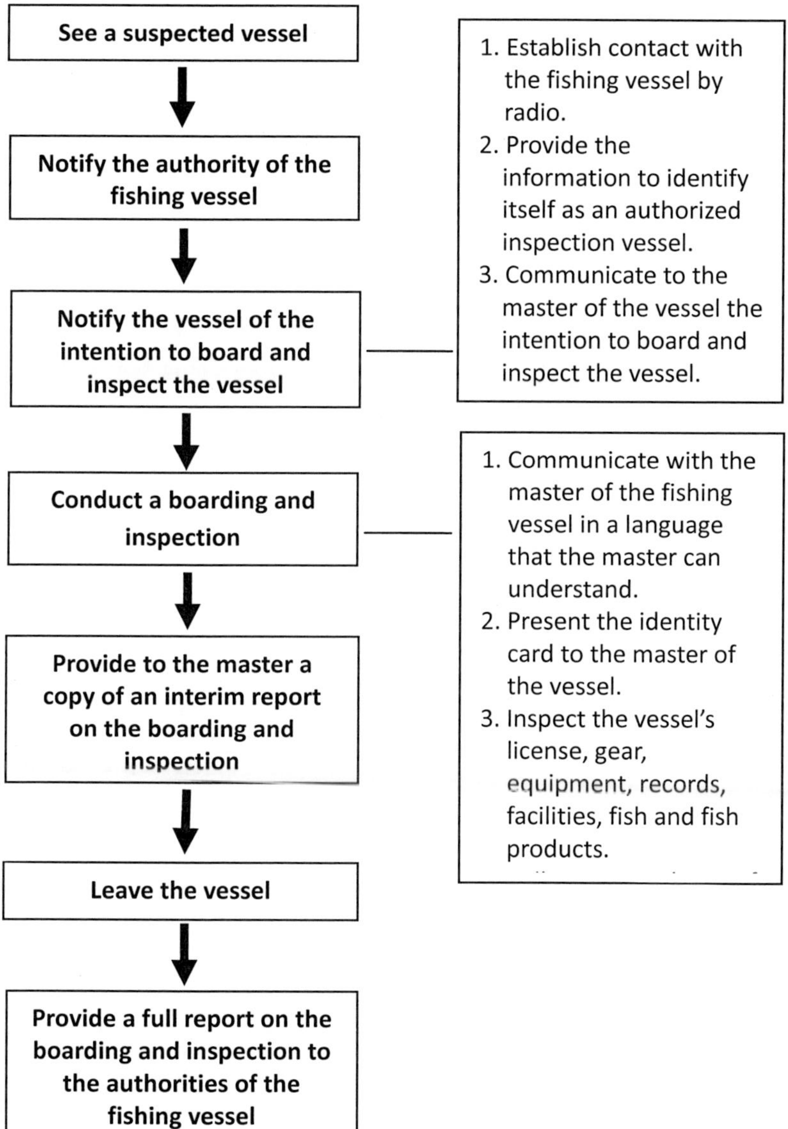

Source: WCPFC Boarding and Inspection Procedure §20-24

B. Fishing Entities' Equivalent to Non-flag State Enforcement in RFMOs

Fishing entities would be regarded as possessing legal personality in the international legal system if the existing contracting parties or members considered the fishing entities to have capability to bear rights and obligations of their legal system and agreed to them becoming members.[75] Therefore, it is possible for fishing entities to participate in enforcement under a regional treaty. However, as the concept of fishing entities mainly stems from the UNFSA, the WCPFC is the only RFMO to allow fishing entities as members and undertake boarding and inspection procedures.[76] However, provisions concerning fishing entities' rights and obligations in the WCPFC Boarding and Inspection Procedure are ambiguous. Article 6 clearly indicates that the procedure applies to fishing entities: "Unless otherwise decided by the Commission, these procedures shall also apply in their entirety as between a Contracting Party and a Fishing Entity, subject to a notification to that effect to the Commission from the Contracting Party concerned." Article 6 does not clearly indicate whether the fishing entity to which it refers is a fishing entity which has already acceded to the WCPFC Convention and become a member of the WCPFC according to the Arrangement for the Participation of Fishing Entities or a fishing entity which is outside the WCPFC regime. If the latter, this Article seems unnecessary because the same regulation can also apply to states outside the regime of WCPFC conservation and management measures; in this circumstance, there is no need to specifically regulate the contracting party and a fishing entity. The reasonable presumption is that it refers to a fishing entity which has acceded to the WCPFC Convention and become a member.

However, the WCPFC Boarding and Inspection Procedure automatically applies to members who accede to the convention; in other words, it is unnecessary to deliver another notification to the commission expressing agreement that the procedures apply to each other. Therefore, Article 6 can be regarded as implying that even as WCPFC members, fishing entities can fall outside the WCPFC Boarding and Inspection Procedure. A possible reason for this stipulation is that boarding and inspection procedures are concerned with infringing a state's sovereignty, and states might not be willing to see their vessels boarded and inspected by a fishing entity. Nevertheless, the purpose of non-flag state enforcement in fisheries within RFMOs is to ensure that the conservation and management

---

[75] Fishing entities' legal status under international law is discussed in Chapter 2 of this book.

[76] The Convention of NPAFC contains provisions concerning non-flag state enforcement but still does not allow fishing entities to become parties or members.

measures can be carried out completely. All WCPFC members, except fishing entities, have the obligation to comply with Procedure, leading to a problematic consequence that whether the WCPFC boarding and inspection regime can be carried out effectively and achieve its aim depends on the willing cooperation of fishing entities. In other words, if a fishing entity delivers a notification of its willingness to comply with the procedure and all other members express willingness to apply the procedure between them and that fishing entity, then the rights and obligations in the Procedure can be established between the state members and the fishing entity. However, if a fishing entity is not willing to comply with the procedure or it delivers such notification but all or some state members are not willing to apply the procedure between them and that fishing entity, then the fishing entity can be legally outside WCPFC Boarding and Inspection Regime. Hence, it is possible that the stipulations concerning notification from fishing entities become a loophole in the WCPFC boarding and inspection regime for either a fishing entity or a state member.

In addition, it implies that Article 6 of the Procedure returns the relation between the fishing entities and contracting parties in the WCPFC boarding and inspection regime to the level of bilateral treaties. This would deviate from the main purpose in establishing RFMOs to bring all actors under conservation and management measures.

C. Bilateral Treaties

States making the non-flag state enforcement measures on the high seas to conserve fishery resources through bilateral treaties was not a new notion, even before the regulations of the UNCLOS, UNFSA and RFMOs. In 1956, Japan and the Soviet Union signed the Japan-Soviet Convention for the Northwest Pacific Fisheries,[77] which allowed Japan and the Soviet Union to board and inspect each other's vessels if there were reasonable grounds to believe the vessels were engaged in activities in violation of the agreement. In 1976, the United States and Soviet Union concluded the Agreement between the Government of the United States of America and the Government of the Union of Soviet Socialist Republics on Mutual Fisheries Relations.[78] Article VIII permitted their vessels to be boarded

[77] *See* Syma A. Ebbin, *A Sea Change: The Exclusive Economic Zone and Governance Institutions for Living Marine Resources* (Dordrecht: Springer, 2005), p. 83; Douglas M. Johnston, *Marine Policy and the Coastal Community: The Impact of the Law of the Sea* (London: Croom Helm Ltd., 1976), pp. 223-224.

[78] The full text can be found at *http://www.state.gov/documents/organization/138873.pdf* (visited on 04/01/2012).

and inspected by each other's duly authorised officers, who could even seize or arrest the vessels and the individuals on board.[79]

After the agreement with the Soviet Union, the United States reached a series of bilateral boarding and inspection agreements with many states beginning in 1976 to ensure fisheries conservation of the high seas of the Atlantic the Agreement between the Government of the United States of America and the Government of the Polish People's Republic Concerning Fisheries off the Coasts of the United States (1976)[80], the Agreement between the Government of the United States of America and the Government of the Republic of China Concerning Fisheries off the Coasts of the United States (1976)[81], the Agreement between the Government of the United States of America and the Government of the People's Republic of Bulgaria Concerning Fisheries off the Coasts of the United States (1976)[82], the Agreement between the Government of the United States of America and the Government of the Socialist Republic of Romania

---

[79] Article VIII of the Agreement states:

"1. Each Party consents to and, to the extent allowable under its own law, will assist and facilitate boardings and inspections of its vessels by duly authorized officers of the other Party for compliance with laws and regulations referred to in Article III. If, upon boarding and inspection of a vessel by a Party's duly authorized officer, such law or regulation is found to have been violated, each Party agrees that it will not object to appropriate enforcement action undertaken pursuant to the laws of that other Party, including seizure and arrest of the vessel and the individuals on board.

2. Each Party shall impose appropriate penalties, in accordance with its laws, for violations of the laws or regulations referred to in Article III. In the case of arrest and seizure of a vessel of a Party by the authorities of the other Party, notification shall be given promptly through diplomatic channels informing the flag state party of the facts and actions taken.

3. Each Party shall release vessels of the other Party and their crews promptly, subject to the posting of reasonable bond or other security.

4. The penalty for violation of a limitation or restriction on the fishing operations of a Party shall be limited to appropriate fines, forfeitures or revocation or suspension of fishing privileges."

[80] The full text is available at *http://untreaty.un.org/unts/144078_158780/3/5/11639.pdf* (visited on 11/12/2011).

[81] The full text is available at *http://www.intfish.net/treaties/bilaterals/texts/chi-usa/chi-usa-1976.pdf* (visited on 28/11/2007).

[82] United Nations, *Treaty Series* (vol. 1134), p. 127.

Concerning Fisheries off the Coasts of the United States (1976)[83], the Agreement between the Government of Spain and the Government of the United States of America Concerning Fisheries off the Coasts of the United States (1977)[84], the Agreement between the Government of the United States of America and the Government of the Republic of Cuba Concerning Fisheries off the Coasts of the United States (1977)[85], the Agreement between the Government of the United States of America and the Government of Mexico Concerning Fisheries off the Coasts of the United States (1977)[86], the Agreement between the Government of the United States of America and the Government of Portugal Concerning Fisheries off the Coasts of the United States (1980)[87], the Agreement between the Government of the United States of America and the Government of Norway Concerning Fisheries off the Coasts of the United States (1981)[88], the Agreement between the Government of the United States of America and the Government of the Republic of Korea Concerning Fisheries off the Coasts of the United States (1982)[89], the Agreement between the Government of the United States of America of the one part and the Government of Denmark and the Home Government of the Faroe Islands of the other part Concerning Fisheries off the Coasts of the United States (1984)[90], the Agreement between the Government of the United States of America and the Government of the Republic of Iceland Concerning Fisheries off the Coasts of the United States (1984)[91],

---

[83] The full text is available at *http://www.intfish.net/treaties/bilaterals/texts/rom-usa/rom-usa-1976.pdf* (visited on 28/11/2007).

[84] The full text is available at *http://www.intfish.net/treaties/bilaterals/texts/spa-usa/spa-usa-1977.pdf* (visited on 28/11/2007).

[85] The full text can be seen in "Cuba—United States: Agreement Concerning Fisheries off the Coasts of the United States," *International Legal Materials*, Vol. 16, No. 3(May 1977), pp. 596–605.

[86] The full text is available at *http://www.intfish.net/treaties/bilaterals/texts/mex-usa/mex-usa-1977.pdf* (visited on 28/11/2007).

[87] The full text is available at *http://faolex.fao.org/docs/texts/por22901.doc* (visited on 12/12/2011).

[88] The full text can be found in:
http://untreaty.un.org/unts/60001_120000/9/21/00017008.pdf
(visited on 12/12/2012).

[89] The full text is available at *http://www.intfish.net/treaties/bilaterals/texts/rok-usa/rok-usa-1982a.pdf* (visited on 28/11/2007).

[90] The full text can be found in:
http://untreaty.un.org/unts/144078_158780/1/1/100.pdf (visited on 12/12/2011).

[91] The full text is available at *http://www.intfish.net/treaties/bilaterals/texts/ice-usa/ice-usa-1984.pdf* (visited on 28/11/2007).

and the Agreement between the Government of the United States of America and the European Economic Community Concerning Fisheries off the Coasts of the United States (1984)[92].

Making a bilateral treaty with other entities or states is not impossible for fishing entities if they or the states regard the fishing entities as possessing the capacity, or the international legal personality, to bear rights and obligations under the international legal system. In fact, non-flag state enforcement on the high seas through bilateral treaties is still based on the consent and authorisation of the flag state. Within bilateral treaties, a fishing vessel of the contracting party could only be boarded and inspected by the other party. It, to a certain degree, preserves more rights for flag states and is much simpler because it involves enforcement only between two parties. However, RFMOs, which have gradually covered almost every high seas area around the world, have provided a forum for states to discuss and decide all existing matters or problems within different areas, so non-flag state enforcement on the high seas is increasingly being developing and carried out within RFMOs. In addition, the concept of fishing entities formally stems from the UNFSA which encourages fishing entities to be brought into RFMOs. Therefore, in addition to having the ability to conclude bilateral treaties concerning enforcement, fishing entities, as the members of RFMOs, are likely to comply with the regulations of non-flag state enforcement under RFMOs.

5. Conclusion

At the bilateral treaty level, there is no impediment for a fishing entity to participate in non-flag state high seas enforcement because international treaties are always concluded based on the parties' willingness. Theoretically, through the UNFSA, the fishing entity can become a member or a cooperative party in RFMOs and further participate in their high seas enforcement schemes. Nevertheless, the WCPFC's Boarding and Inspection Procedure excludes fishing entity from holding a position equivalent to that of a normal non-flag state obliged to conform to the procedure. In contrast, by the delivery of notification, a fishing entity can decide whether it is inside or outside the procedure, and the state members can decide whether to apply the Procedure between themselves and the fishing entity. Consequently, the fishing entity's role in non-flag enforcement is still based on the nature of bilateral treaties; the only

[92] The full text is available at *http://faolex.fao.org/docs/texts/bi-11049.doc* (visited on 12/12/2011).

difference is that there is an intermediary, the RFMO. The following chapter further discusses the problematic consequences of enforcement by fishing entities on the high seas.

CHAPTER SIX

THE PROBLEMATIC CONSEQUENCES OF ENFORCEMENT BY FISHING ENTITIES ON THE HIGH SEAS

1. Introduction

A fishing entity, as an equivalent to a flag state, should not to encounter many problems in high seas enforcement, but due to the nature of non-flag state enforcement on the high seas as a supplement to traditional flag state enforcement, fishing entities might face some uncertainties in this area. At present, the legal basis for non-flag state enforcement on the high seas comes either from bilateral or multilateral treaties. Under bilateral treaties, both parties can board and inspect each other's vessels. The advantage of bilateral treaties is that they are simpler and can be concluded more easily. On the other hand, through multilateral treaties, more states are involved in the boarding and inspection procedure, increasing the effectiveness of the fisheries sustainable development. Under a multilateral treaty, all states can board and inspect other parties' vessels and must allow their own vessels to be boarded and inspected by others. However, due to the larger number of states which might have different interests in their fisheries policy, it becomes complicated to negotiate, concluded and implement multilateral treaties. Being a party to bilateral treaties concerning high seas enforcement establishes a clear legal status for fishing entities as a non-flag state. Thus, at the bilateral level, fishing entities can participate in non-flag state enforcement on the high seas to conserve and manage fisheries resources. However, at the multilateral level, such as multilateral conventions or RFMOs, states are usually the subject of regulations; therefore, enforcement by fishing entities poses an issue in the international law of the sea which is difficult to resolve.

In addition, neither the UNCLOS nor the UNFSA allow fishing entities to accede; only some RFMOs allow fishing entities to become members. However, UNCLOS and the UNFSA play significant roles in fisheries conservation and management on the high seas. In particular, the UNFSA establishes a high seas boarding and inspection procedure as a temporary measure for RFMOs which had not developed such a procedure within two years of its adoption. Fishing entities, as members of the RFMOs, face not only the problem of compliance with the UNFSA but also the uncertainty of their rights and obligations in RFMOs regarding non-flag state high seas enforcement. These uncertainties and problems that fishing entities, as equivalent to a non-flag state, might encounter are discussed in this chapter. Furthermore, the WCPFC is the only RFMO which both allows fishing entities to be members and has adopted boarding and inspection procedure according to the UNFSA. Therefore, this chapter also discusses the uncertainty that fishing entities might face in the WCPFC Boarding and Inspection Procedure.

2. Weakening the Effectiveness of Multi-protection in Conservation and Management Systems

Articles 63 and 64 in UNCLOS stipulate that relevant states should cooperate directly or indirectly through RFMOs in establishing conservation and management measures for straddling and highly migratory fish stocks. Articles 116–119 reiterate the importance of cooperation among states and provide the principles for conservation and management measures for living resources in the high seas. Based on those principles, the 1995 UNFSA was adopted to establish more detailed provisions for the implementation of UNCLOS.[1] Article 4 of the UNFSA states that "[n]othing in this Agreement shall prejudice the rights, jurisdiction and duties of States under the Convention. This Agreement shall be interpreted and applied in the context of and in a manner consistent with the Convention." Therefore, although the UNFSA is a separate treaty from UNCLOS, the role of the UNFSA is to supplement the UNCLOS. The conservation and management measures adopted under the UNFSA should not prejudice the related provisions in the UNCLOS.[2]

As the implementation of UNCLOS, UNFSA uses RFMOs as a major method to strengthen and carry out its conservation and management measures. Articles 8(1) and (3) state that the related states should cooperate through RFMOs and have an obligation to follow the conservation and

[1] Article 2 of the UNFSA states that the purpose of the agreement is to "ensure the long-term conservation and sustainable use of straddling fish stocks and highly migratory fish stocks through effective implementation of the relevant provisions of the Convention."

[2] *See* Article 7 of the UNFSA.

management measures set by RFMOs. Article 8(5) further states that, if there is no RFMO to regulate conservation and management, "relevant coastal States and States fishing on the high seas for such stock in the subregion or region shall cooperate to establish such an organisation or enter into other appropriate arrangements to ensure conservation and management of such stock and shall participate in the work of the organisation or arrangement." Articles 9–14 of the UNFSA lay out the functions and issues that RFMOs should observe in their conservation and management measures.

As parties to the UNCLOS, UNFSA or any treaties that create RFMOs, states create a situation in high seas fisheries conservation and management which can be called uni-protection. On the other hand, UNFSA regulations concerning UNCLOS and RFMOs create a triangle system among these entities. Compared with uni-protection, states bound by these three regulations create the situation of multi-protection, increase the strength of fisheries conservation and management on the high seas. Within the multi-protection system, the UNCLOS provides a framework for high seas fisheries resources conservation and management. Based on UNCLOS, the UNFSA further focuses on the sustainable development of straddling and highly migratory fish stocks. RFMOs carry out the principles of the conservation and management measures in the UNCLOS and UNFSA and establish concrete procedures or measures to implement this convention and agreement. Consequently, the UNCLOS, UNFSA and RFMOs are interdependent and mutually reinforce the conservation and management measures, building a tight network for the sustainable utilisation of high seas fisheries resources.[3] Within this network, the UNFSA stands as the centre connecting the UNCLOS and RFMOs. Although the UNFSA has provisions concerning fishing entities, it does not allow them to join.[4] On the other hand, the agreement does not exclude but even encourages fishing entities to participate in RFMOs[5] in order to bring fishing entities into conservation and management networks.

---

[3] *See* Peter G.G. Davies and Catherine Redgwell, "The International Legal Regulation of Straddling Fish Stocks," *The British Yearbook of International Law*, Vol. 67(1996), pp. 270; 272.

[4] Article 1(3) states: "This Agreement applies *mutatis mutandis* to other fishing entities whose vessels fish on the high seas." However, Article 39 concerning accession does not allow fishing entities to join. Regarding fishing entities' status in the UNFSA, please *see* Chapter 2 of this book.

[5] Article 17(3) reads: "States which are members of RFMOs should individually or jointly request fishing entities to cooperate with RFMOs and such fishing entities shall enjoy benefits from participation in the fishery commensurate with their

Any state would be bound by the agreement if it became a state party to the UNFSA, regardless of whether it also participates in a RFMO. If fishing entities become members of a RFMO, it is absolutely clear that they are directly bound by the RFMO's conservation and management measures and so indirectly follow the related UNFSA regulations. However, if a fishing entity does not or refuses to participate in RFMOs, then it is not be bound by the regulations of either RFMOs or the UNFSA. Consequently, establishing the relations between RFMOs and fishing entities becomes significant to bring fishing entities into the conservation and management system.

The UNCLOS, UNFSA and RFMO regulations can be regarded as the filters through which the more states pass, the more stable the conservation and management regime becomes. In other words, if more states are parties to all three sets of regulations, more states are bound by the multi-protection system. In this context, the best situation is for all the actors who have vessels conducting fishing on the high seas to be brought into the entire triangle system, rather than just one or two of the regulatory systems. However, the lack of opportunity for fishing entities to join the UNFSA and UNCLOS places stress on RFMOs to play the crucial role of deciding whether fishing entities are inside or outside conservation and management regimes. The lack of opportunity also raises the key question of whether the UNFSA and RFMOs' measures can be fully enforced on the high seas. Accordingly, RFMOs become the only international conservation and management regimes which may have binding effect on fishing entities. The effectiveness of the conservation and management measures, therefore, is reduced from multi-protection to uni-protection.[6]

3. Uncertainty of the Relations among UNFSA, RFMOs and Fishing Entities

The relation between the UNFSA and RFMOs can be seen in Article 8 of the UNFSA. Article 8(1) states that coastal states and states fishing on the high seas should cooperate on straddling and highly migratory fish stocks

---

commitment to comply with conservation and management measures in respect of the stocks."

[6] *See* also Rosemary Rayfuse, "The Interrelationship between the Global Instruments of International Fisheries Law," in Ellen Hey, ed., *Developments in International Fisheries Law* (Hague: Kluwer Law International, 1999), pp. 153–155.

directly or indirectly through RFMOs.[7] Article 8(5) further stipulates that those states should establish a RFMO if no such organisation exists.[8] Therefore, the UNFSA provides that, if a RFMO existed before the adoption of the UNFSA, state parties may cooperate through that RFMO; if there was no RFMO, states parties should establish one.

Article 21(2) of the UNFSA stipulates that states should through RFMOs establish a boarding and inspection procedure. Article 21(3) of the UNFSA states that, if a RFMO does not establish its own boarding and inspection procedures within two years of the adoption of the UNFSA, the UNFSA Boarding and Inspection Procedure in Article 22 should apply automatically as between UNFSA parties.[9] Pursuant to Article 21 of the UNFSA, the RFMOs are encouraged and expected to establish their own boarding and inspection procedures. Nevertheless, the UNFSA does not explicitly indicate if the RFMOs which correspond with those in Article 21(3) adopting the UNFSA boarding and inspection procedures are the new organisations established after the adoption of the UNFSA or the ones that existed before 1995. It can be understood that RFMOs established on the basis of the UNFSA may apply Article 21(3) of the UNFSA and adopt its boarding and inspection procedures.[10] However, RFMOs set up

[7] The full text of Article 8(1) reads: "Coastal States and States fishing on the high seas shall, in accordance with the Convention, pursue cooperation in relation to straddling fish stocks and highly migratory fish stocks either directly or through appropriate subregional or regional fisheries management organizations or arrangements, taking into account the specific characteristics of the subregion or region, to ensure effective conservation and management of such stocks."

[8] Article 8(5) states: "Where there is no subregional or regional fisheries management organization or arrangement to establish conservation and management measures for a particular straddling fish stock or highly migratory fish stock, relevant coastal States and States fishing on the high seas for such stock in the subregion or region shall cooperate to establish such an organization or enter into other appropriate arrangements to ensure conservation and management of such stock and shall participate in the work of the organization or arrangement."

[9] Article 21(3) reads: "If, within two years of the adoption of this Agreement, any organization or arrangement has not established such procedures, boarding and inspection pursuant to paragraph 1, as well as any subsequent enforcement action, shall, pending the establishment of such procedures, be conducted in accordance with this article and the basic procedures set out in article 22."

[10] At present, the WCPFC is the only RFMO which complies with Article 21(3) of the UNFSA, through wording in Article 26(2) of the WCPFC Convention that reads: "If, within two years of the entry into force of this Convention, the Commission is not able to agree on such procedures, or on an alternative mechanism which effectively discharges the obligations of the members of the Commission under the Agreement and this Convention to ensure compliance with

separately before the UNFSA do not have the obligation to cooperate with Article 21(3), so the UNFSA needs to establish another special link with RFMOs.[11] Even if an RFMO was formed after the UNFSA, it could be possible to refuse to implement Article 21(3) of the UNFSA. The WCPFC reached a gentleman's agreement to not implement the UNFSA boarding and inspection rules before concluding its own procedures.[12] This uncertainty in relations between RFMOs and the UNFSA can cause confusion in the application of Article 21(3) of the UNFSA.

In addition, under Article 21(3), a member of a RFMO, whether formed before or after, which is also a state party of the UNFSA may conduct boarding and inspection between it and other state parties. Fishing entities which cannot become state parties of the UNFSA face two possible situations. If a fishing entity is a member of a RFMO, it raises doubt whether Article 22 of the UNFSA can apply to the fishing entity. Pursuant to Article 21(2) of the UNFSA, the RFMO should adopt its own boarding and inspection procedures, which should be consistent with Article 22 of the UNFSA. As a member of the RFMO, the fishing entity should follow the regulations adopted by the RFMO; therefore, the boarding and inspection procedures provided in Article 22 of the UNFSA might automatically apply to the fishing entity, which thus is indirectly bound to the UNFSA.

In the other situation, if a fishing entity does not participate in a RFMO, it is not be bound by the boarding and inspection procedures of the UNFSA. In this circumstance, the RFMO complies with the UNFSA boarding and inspection procedures, but the fishing entity is excluded from the procedures, as well as the conservation and management regime on the high seas. Hence, while implementing boarding and inspection procedures among members of a RFMO, the fishing entity has no warrant to board and inspect the vessels of other RFMO members; on the other hand, the

---

the conservation and management measures established by the Commission, articles 21 and 22 of the Agreement shall be applied, subject to paragraph 3, as if they were part of this Convention and boarding and inspection of fishing vessels in the Convention Area, as well as any subsequent enforcement action, shall be conducted in accordance with the procedures set out therein and such additional practical procedures as the Commission may decide are necessary for the implementation of articles 21 and 22 of the Agreement."

[11] Although RFMOs members that are also parties of the UNFSA are bound by the agreement, the RFMOs themselves, as the subjects of international law, still need a special link to the UNFSA.

[12] Douglas Guilfoyle, *Shipping Interdiction and the Law of the Sea* (New York: Cambridge University Press, 2009), p. 110.

vessels of the fishing entity could also escape being boarded and inspected on the high seas, increasing the possibilities for the vessels of the fishing entity to violate the conservation and management measures of the RFMO. This would further weaken the effectiveness of the boarding and inspection procedures of the UNFSA.

4. Confusing Legal Status in the WCPFC Enforcement Procedures

Pursuant to Article 26[13] of the WCPFC Convention, the WCPFC Boarding and Inspection Procedure[14] was adopted in 2006 and applies on the high seas within the convention area. Theoretically, members discussed the boarding and inspection procedure during the WCPFC regular meeting, so each member should equally cooperate and participate in this procedure. It means that a member of the WCPFC possesses equal right to board and inspect the vessels of all other members but also has the obligation to ensure that its vessels may be boarded and inspected on the high seas within the convention area. However, it seems that this procedure does not distribute rights and obligations equally but distinguishes between the rights and obligations by contracting parties and the members.

Articles 2(c) and 2(d) of the WCPFC Boarding and Inspection Procedure separately define "Authorities of the Inspection Vessel" and "Authorities of the Fishing Vessel." The former means the authorities of the contracting party of the commission under whose jurisdiction the inspection vessel is operating. The latter means the authorities of the member of the commission under whose jurisdiction the fishing vessel is operating. In the WCPFC Boarding and Inspection Procedure, the subjects which can be inferred to have the right to board and inspect other members' vessels are defined by the term "contracting party," and the subjects which can be inferred to have obligations to allow being boarded and inspected

---

[13] Article 26(1) states: "For the purposes of ensuring compliance with conservation and management measures, the Commission shall establish procedures for boarding and inspection of fishing vessels on the high seas in the Convention Area. All vessels used for boarding and inspection of fishing vessels on the high seas in the Convention Area shall be clearly marked and identifiable as being on government service and authorized to undertake high seas boarding and inspection in accordance with this Convention."

[14] The full text of the procedure can be found on the official WCPFC website, *http://www.wcpfc.int/doc/cmm-2006-08/western-and-central-pacific-fisheries-commission-boarding-and-inspection-procedures* (visited on 21/05/2012). The details of this procedure are discussed in Chapter 5 of this book.

are described by the term "member of the commission." The result is that the contracting parties of the WCPFC Convention and the members of the commission might have different legal statuses under the WCPFC Boarding and Inspection Procedure.

With respect to the subjects that can be inferred to have the rights of boarding and inspection, Article 5 of the WCPFC Boarding and Inspection Procedure states that "[e]ach Contracting Party may, subject to the provisions of these procedures, carry out boarding and inspection on the high seas of fishing vessels engaged in or reported to have engaged in a fishery regulated pursuant to the Convention." Article 13 regulates the documents and notifications that the inspectors and inspection vessels should provide: "Each Contracting Party that intends to carry out boarding and inspection activities pursuant to these procedures shall so notify the Commission, through the Executive Director, and shall provide the following …." Article 15 further stipulates that authorised inspection vessels and inspectors notified by contracting parties pursuant to paragraph 13 shall be included in the commission register once the executive director confirms that they meet the requirements of that paragraph. Article 16 instructs that, in order to increase the effectiveness of the procedure, "[c]ontracting Parties may identify opportunities to place authorised inspectors on inspection vessels of another Contracting Party. … Contracting Parties should seek to conclude bilateral arrangements … or otherwise facilitate communication and coordination between them for the purpose of implementing these procedures." Article 40 requires contracting parties whose vessels conduct boarding and inspection under the WCPFC Boarding and Inspection Procedure to report annually to the commission about the enforcement activities they carried out. In addition, the contracting parties should take responsibility for any damage or loss caused by their actions while implementing the WCPFC Boarding and Inspection Procedure.[15] Finally, Article 47 provides a way that the contracting parties may seek to promote optimum use of the authorised inspection vessels and authorised inspectors when applying these procedures.[16]

[15] Article 45 of the WCPFC Boarding and Inspection Procedure reads: "Contracting Parties shall be liable for damage or loss attributable to their action in implementing these procedures when such action is unlawful or exceeds that reasonably required in the light of available information."

[16] The full text of Article 47 of the WCPFC Boarding and Inspection Procedure reads: "The Commission shall keep under continuous review the implementation and operation of these procedures, including review of annual reports relating to these procedures provided by Members. In applying these procedures, Contracting

With regard to the subjects that can be inferred to have obligations under the WCPFC Boarding and Inspection Procedure, Article 7 states that "[e]ach Member of the Commission shall ensure that vessels flying its flag accept boarding and inspection by authorized inspectors in accordance with these procedures." Article 27 reads: "The authorities of the fishing vessel, unless generally accepted international regulations, procedures and practices relating to safety at sea make it necessary to delay the boarding and inspection, shall direct the master to accept the boarding and inspection. If the master does not comply with such direction, the Member shall suspend the vessel's authorization to fish and order the vessel to return immediately to port. The Member shall immediately notify the authorities of the inspection vessel and the Commission of the action it has taken in these circumstances."[17]

As a result, although the WCPFC Boarding and Inspection Procedure does not explicitly state that only contracting parties have the right to board and inspect other members' vessels, its language seems to imply so. Interestingly, the procedure was initially provided by the chairman under the title of the Draft Procedures for Boarding and Inspection Pursuant to the WCPFC Convention[18] in Working Group III of the fourth session of

---

Parties may seek to promote optimum use of the authorized inspection vessels and authorized inspectors by:

a. identifying priorities by area and/or by fishery for boarding and inspections pursuant to these procedures;
b. ensuring that boarding and inspection on the high seas is fully integrated with the other monitoring, compliance and surveillance tools available pursuant to the Convention;
c. ensuring non-discriminatory distribution of boarding and inspections on the high seas among fishing vessels of Members of the Commission without compromising the opportunity of Contracting Parties to investigate possible serious violations; and
d. taking into account high seas enforcement resources assigned by Members of the Commission to monitor and ensure compliance by their own fishing vessels, particularly for small boat fisheries whose operations extend onto the high seas in areas adjacent to waters under their jurisdiction."

[17] The articles above mentioned concerning the rights and obligations in the WCPFC Boarding and Inspection Procedure are some examples. The distinction between contracting parties and members can be seen throughout the procedure.

[18] The full text can be found at *http://www.wcpfc.int/doc/wcpfcprepconwp14/draft-procedures-boarding-and-inspection-pursuant-convention* (visited on 10/08/2012).

the WCPFC Preparatory Conference (PrepCon IV).[19] Then, in light of discussions during PrepCon IV, the chairman prepared the Revised Draft Procedures for Boarding and Inspection Pursuant to the WCPFC Convention[20] in the fifth session of the WCPFC Preparatory Conference (PrepCon V) on 10 September 2003.[21] In these two documents, the subject of the Procedures was drafted as "[Contracting Party] [Member of the Commission]." It is obvious that the participants had not decided on the exact subject. In addition, in the first session of the Technical and Compliance Committee (TCC1)[22] in 2005, the issue of the Boarding and Inspection Procedure was discussed, and the report of the TCC1 mentions that one of the key outstanding issues which requires resolution is whether the authority to board and inspect on the High Seas should be available to all members of the commission or limited only to contracting parties.[23] Under this question, a footnote states:

> The significance of this issue lies in the fact that, in a few cases, the Convention draws a subtle distinction between "Contracting Parties" and "Members of the Commission." "Contracting Parties" are States or entities entitled to ratify or accede to the Convention. "Members of the Commission" include all "Contracting Parties" and others entitled to membership of the Commission. Members may therefore include States, Regional Integration Organizations and Fishing Entities. … Significantly, Article 26 of the Convention dealing with boarding and inspection powers comes under Part VI of the Convention entitled "Compliance and Enforcement." It is worth noting that all the substantive provisions under Part VI apply to "Members of the Commission." Thus, it can be concluded

---

[19] Further information about PrepCon VI can be found in the Summary Report by the Chairman of Working Group III to the Fourth Session of the Conference, available at *http://www.wcpfc.int/doc/wcpfcprepcon26/summary-report-chairman-working-group-iii-fourth-session-conference* (visited on 10/08/2012).

[20] The full text can be found at *http://www.wcpfc.int/doc/wcpfcprepconwp14-rev-1/draft-procedures-boarding-and-inspection-pursuant-convention* (visited on 10/08/2012).

[21] Further information about PrepCon V can be found in the Summary Report by the Chair of Working Group III to the Fifth Session of the Preparatory Conference, available at *http://www.wcpfc.int/doc/wcpfcprepcon33/summary-report-chair-working-group-iii-fifth-session-preparatory-conference* (visited on 10/08/2012).

[22] *See* the 1st Regular Session of the Technical and Compliance Committee on 5–9 December 2005, available at *http://www.wcpfc.int/meetings/2005/1st-regular-session-technical-and-compliance-committee* (visited on 12/08/2012).

[23] *See* WCPFC-TCC1-15, Boarding and Inspection Procedures, para 8, available at *http://www.wcpfc.int/doc/wcpfc-tcc1-2005-15/boarding-and-inspection-procedures-53k* (visited on 12/08/2012).

> that the boarding and inspection powers envisaged under Article 26 of the Convention are applicable to all Members of the Commission.[24]

It seemed as if the subject of the procedure likely would be members of the commission; if so, a fishing entity could unambiguously be the subject of the procedure. On 10 March 2006, the TCC2[25] concluded the WCPFC Commission Boarding and Inspection Procedures, which defines the subject of the Procedure as "members of the commission" in a footnote stating:

> The use of the term "Member" throughout the text is without prejudice to the position of any delegation with respect to the final disposition of the provisions of these procedures regarding the participation of Fishing Entities.[26]

The report of the TCC2 also states that the use of the term "member" throughout the text should be considered provisional, pending final resolution of this matter.[27] However, in the third annual meeting of the WCPFC in December 2006, the final version of the Boarding and Inspection Procedure was adopted, making the subject of the Procedure "contracting parties," not "members of the commission."

Fishing entities cannot join the WCPFC Convention through accession[28] but can express their consent to be bound by the convention

---

[24] *Ibid.*, fn 3.

[25] *See* the Second Regular Session of the Technical and Compliance Committee on 28 September-3 October 2006, available at *http://www.wcpfc.int/meetings/2006/2nd-regular-session-technical-and-compliance-committee* (visited on 13/08/2012).

[26] *See* WCPFC-TCC2-12, High Seas Boarding and Inspection Procedures and Attachments, fn 1 available at *http://www.wcpfc.int/doc/wcpfc-tcc2-2006-12/high-seas-boarding-and-inspection-procedures-attachments* (visited on 13/08/2012).

[27] *See* WCPFC-TCC2-2006- Summary, Summary Record with Attachments, para 71, p. 14, available at *http://www.wcpfc.int/doc/wcpfc-tcc2-2006-summary/summary-record-with-attachments* (visited on 13/08/2012).

[28] Article 35 of the WCPFC Convention stipulates the ways for other states or entities may accede to it. Article 35(1) stipulates that the states referred to in Article 34(1) and any entity referred to in Article 305(1)(c),(d) and (e) of UNCLOS which are in the convention area may join. The states mentioned in Article 34(1) are Australia, Canada, China, the Cook Islands, Federated States of Micronesia, Fiji Islands, France, Indonesia, Japan, Republic of Kiribati, Republic of the Marshall Islands, Republic of Nauru, New Zealand, Niue, Republic of Palau, Independent State of Papua New Guinea, Republic of the Philippines, Republic of Korea, Independent State of Samoa, Solomon Islands, Kingdom of Tonga, Tuvalu,

and become WCPFC members. Paragraph 1 of Annex I of the WCPFC Convention provides a method for fishing entities to join: "After the entry into force of this Convention, any fishing entity whose vessels fish for highly migratory fish stocks in the Convention Area, may, by a written instrument delivered to the depositary, agree to be bound by the regime established by this Convention. Such agreement shall become effective thirty days following the delivery of the instrument. Any such fishing entity may withdraw such agreement by written notification addressed to the depositary. The withdrawal shall take effect one year after the date of receipt of the notification, unless the notification specifies a later date." In addition, Paragraph 2 of Annex I states that fishing entities who join the WCPFC Convention through Paragraph 1 of the Annex should participate in the work of the commission, including decision-making, and should comply with the obligations under this convention; in other words, these fishing entities should also become members of the WCPFC.

Moreover, Paragraph 2 of Annex I further stipulates that references to the commission or members of the commission include "such fishing entity as well as Contracting Parties." Accordingly, fishing entities can be members of the WCPFC but are still not the same as contracting parties. Then, fishing entities, as members of the WCPFC, certainly must ensure that their vessels may be boarded and inspected by other contracting parties of the WCPFC Convention. However, it remains a question of whether fishing entities can be regarded as equivalent to contracting parties or merely members in the WCPFC Boarding and Inspection Procedure. Thus, it is unclear whether fishing entities can conduct boarding and inspection of other members' vessels according to the WCPFC Boarding and Inspection Procedure. Although Article 6 of the WCPFC Boarding and Inspection Procedure states that it also applies to fishing entities, it does not clarify the rights and obligations of fishing entities.

The Technical and Compliance Committee in its third regular session in 2007 (TCC 3) adopted the High Seas Boarding and Inspection Procedure

---

the United Kingdom of Great Britain and Northern Ireland in respect of Pitcairn, Henderson, Ducie and Oeno Islands, the United States of America and Republic of Vanuatu. In addition, Article 35(2) of the WCPFC Convention states: "After the entry into force of this Convention, the Contracting Parties may, by consensus, invite other States and regional economic integration organizations, whose nationals and fishing vessels wish to conduct fishing for highly migratory fish stocks in the Convention Area to accede to this Convention."

(Revised)[29](the revised procedure). Paragraph 8 further indicates that the contracting parties in Article 13 of the WCPFC Boarding and Inspection Procedure should provide details about the vessel and the names of the authorities responsible for boarding and inspection. Paragraph 8 includes a footnote on the term "contracting parties" which reiterates Article 6 of the WCPFC Boarding and Inspection Procedure. This footnote further explains that "it follows that when a Contracting Party notifies the Commission that the High Seas Boarding and Inspection Procedures apply in their entirety as between that Contracting Party and a Fishing Entity, the operational guidelines for implementing the Procedures will also apply to the Fishing Entity concerned." This suggests that fishing entities might have equality status if a contracting party has made such a notification.

Neither the WCPFC Boarding and Inspection Procedure nor the revised procedure clarifies fishing entities' legal status. Neither clearly indicates whether the fishing entity to which it refers is a fishing entity which has already acceded to the WCPFC Convention and become a member of WCPFC according to the Arrangement for the Participation of Fishing Entities or a fishing entity which remains outside the WCPFC regime. However, as it is unnecessary for fishing entities to deliver another notification to the commission expressing their agreement to apply the procedure to one another, the possible inference is that fishing entities can go beyond their status as members of WCPFC and outside the WCPFC boarding and inspection regime, which implies that contracting parties and fishing entities can board and inspect each other's vessels only through bilateral treaties.[30]

This ambiguity causes uncertainty in fishing entities' status in the WCPFC Boarding and Inspection Procedure. Furthermore, states cooperate with each other to enforce fishery resources conservation and management measures through boarding and inspection at the multilateral level, *i.e.* through RFMOs, which is more far-reaching than the bilateral level. The effectiveness of the multilateral cooperation might be decreased if the relation between fishing entities and states is returned to the level of bilateral treaties, rather than bringing them all into a common regime. Doing so might make whether conservation and management measures can be carried out dependent upon the willingness of fishing entities. This practice would deviate from the main purpose of establishing RFMOs: to

[29] WCPFC-TCC3-2007-11, in the Third Regular Session of WCPFC Technical and Compliance Committee is available on the official WCPFC website, *http://www.wcpfc.int/doc/wcpfc-tcc3-2007-11/high-seas-boarding-and-inspection-procedures-rev-1* (visited on 05/04/2012).

[30] The related inference is discussed in Chapter 5 of this book.

bring all actors under conservation and management measures if a fishing entity would like to escape from monitoring.[31]

The other possible interpretation is that fishing entities have no choice but to allow their vessels to be boarded and inspected but themselves may not board and inspect others' vessels. However, this interpretation might push fishing entities away from compliance with the boarding and inspection procedure, even those adopted by RFMOs, if they have obligations but no rights in the procedure. It might also influence their willingness to participate in the RFMOs, leading again to deviations from the main objective of RFMOs to bring all actors into the conservation and management regimes.

5. Conclusion

International fisheries law and RFMOs' regulations encourage and expect that fishing entities will be bound by them like states in order to effectively carry out conservation and management measures. However, under present regulations, RFMOs' agreements and documents are the only regulations by which a fishing entity might be bound. In addition, with respect to high seas enforcement, the UNFSA expects that, through the UNFSA's or RFMOs' boarding and inspection procedures, states can board and inspect each other's vessels at the multilateral level. Fishing entities, though, cannot become state parties to the UNFSA and thus are excluded from such high seas enforcement, whereas according to the UNFSA, state parties may conduct boarding and inspections on each other's vessels. Although the WCPFC has adopted its own boarding and inspection procedure, a fishing entity and other states may board and inspect each other's vessels only through notification from both sides. Therefore, fishing entities still may not conduct high seas boarding and inspections on the multilateral level but only on the bilateral. These problems facing fishing entities in regulations regarding high seas enforcement might affect the effectiveness of conservation and management measures on the high seas. In the following chapters, we consider the example of Taiwan, as the only fishing entity so far to participate in RFMOs, and its status in high seas enforcement under various related regimes.

[31] Also *see* Chapter 5 of this book.

PART III:

PRACTICE WITHIN REGIONAL FISHERIES ORGANISATIONS— THE EXAMPLE OF TAIWAN

CHAPTER SEVEN

MEMBERSHIP

1. Introduction

Taiwan is a major distant-water fishing nation with respect to straddling and highly migratory stocks. Hence, it is significant that Taiwan can make a great contribution to the whole conservation and management system. Bringing this strong fishing actor into a controlled regime is necessary. The UNFSA created the term "fishing entity" for Taiwan, giving Taiwan a legal basis in the regime of RFMOs so that their conservation and management measures could be implemented effectively. At present, only Taiwan participates in RFMOs under the identity of fishing entity. This unusual situation stems from the long-running dispute between Taiwan and China, which tries to prevent Taiwan from being directly or indirectly recognised as a state in the international community. The purpose of this chapter is to discuss Taiwan's participation as a fishing entity in RFMOs. This chapter first reviews the background of Taiwan's participation in RFMOs as a fishing entity, functioning as but not possessing the status of a state. This chapter then considers the linkage between Taiwan and the 1995 agreement and finally the process of Taiwan's participation as a member in the RFMOs which comprise the WCPFC, IATTC and ISC.

2. Taiwan's Special Status in International Law—The Reason to be a Fishing Entity

2.1 Background

During the Ching Dynasty[1], Taiwan was part of China until the conclusion of Treaty of Shimonoseki[2] in 1895 which ended a war between China and

[1] The Ching Dynasty was the last monarchy in China and lasted from 1644 to 1912.
[2] The full text of the treaty can be found at *http://china.usc.edu/(S(u0o4uo4552nc4b45rm1khibm)A(4BUbwWp7zQEkAAAAZDlkMDBjYTctYjM1NC00ODg1LWE5YjgtYTg4NGE0ZDc0ODU4BJSk2ve3v12T5hS*

Japan. China lost the war and, in the Treaty of Shimonoseki, agreed to cede territories, including Taiwan, Penghu Islands and the surrounding small islands, to Japan. Under the treaty, Taiwan became a Japanese colony until 1945 when Japanese rule was ended by politics and war, not the Treaty of Shimonoseki itself.[3]

Meanwhile, the Ching Dynasty was overthrown on 10 October 1911 by the Kuo Min Tang Party (the Nationalist Party, the KMT) in a revolution.[4] The Republic of China (ROC), which was the successor to the Ching Dynasty and ended more than two thousand years monarchy in China, was established on 1[t] January 1912. After the Ching Dynasty was overturned, various warlords caused civil wars. To control effectively the whole territory of China, the ROC became involved in those wars. Furthermore, the Communist Party was established in July 1921.[5] Against this background, war between China and Japan broke out in 1937 and became the Far East battlefield of the Second World War after Japan joined the Axis in 1940 and China joined the Allies in 1941.[6] The war was ended in 1945 with the surrender of Japan; afterwards, the ROC was delegated by the Allies to take over Taiwan. At that time, the ROC was an original member state and permanent member of UN Security Council in the UN.[7] However, although China's external war was ended, its civil war with the Communist Party did not cease. The Communist Party established the

---

KPbdqc1b-n8A1))/ShowArticle.aspx?articleID=405&AspxAutoDetectCookieSupport=1 (visited on 01/06/2012).

[3] *See* Denny Roy, *Taiwan: A Political History* (New York: Cornell University Press, 2003), pp. 32–54.

[4] At that time, it was still a revolutionary group; it was formed with the name of KMT in August 1912. *See* Jonathan Fenby, *Generalissimo: Chiang Kai-Shek and the China He Lost* (London: Free Press, 2005), p. 30; p. 35.

[5] Information about the Communist Party of China can be seen at *http://www.chinatoday.com/org/cpc/* (visited on 01/06/2012).

[6] Japan invaded China in 1931 and declared the Great East Asia Coprosperity Sphere in 1940.

[7] *See* the official UN website, *http://www.un.org/en/aboutun/history/index.shtml* (visited on 01/06/2012); and *http://www.un.org/sc/members.asp* (visited on 01/06/2012).

People's Republic of China (PRC)[8] on 1 October 1949, and on 7 December, the ROC military withdrew from mainland China to Taiwan.[9]

2.2 Loss of Membership in the United Nations

Since the establishment of the PRC, it has followed its One China Policy which claims that there is only one China, the PRC is the only legitimate representative of China and Taiwan is a part of China.[10] At the request of 17 members[11] during the 26th Session of the UN General Assembly on 15 July 1971, the question of the restoration of the lawful rights of the People's Republic of China in the United Nations was placed on the provisional agenda. These 17 members stated that China was a founding member of the UN and a permanent member of the Security Council and that the PRC was the only representative of China.[12]

On 25 September 1971, draft resolution A/L.630[13] was submitted to the General Assembly by 23 members (including the previous 17 members). The draft resolution A/L.630 requested that the General Assembly restore to the PRC all its rights and expel the representatives of Chiang Kai-shek, who was the leader of the ROC.[14] Another draft resolution A/L.632

---

[8] The PRC divides China into 23 provinces (including Taiwan although it has never been actually controlled by the PRC), 5 autonomous regions, 4 municipalities directly under the jurisdiction of the Central Government and 2 special administrative divisions (Hong Kong and Macao).

[9] *See* Roy, *supra note* 3, pp. 55-75.

[10] *See* the official website of the PRC government, *http://www.gov.cn/test/2005-07/29/content_18293.htm* (visited on 01/06/2012). The One China Policy is the PRC's core discourse in international forum touching upon the Taiwan issue. On 14 March 2005, the PRC adopted the Anti-Secession Law which codified the "One China Policy." The full text of the Anti-Secession Law is available at *http://english.peopledaily.com.cn/200503/14/eng20050314_176746.html* (visited on 01/06/2012).

[11] These members are Albania, Algeria, Cuba, Democratic Republic of Yemen, Guinea, Iraq, Mali, Mauritania, the People's Republic of the Congo, Romania, Somalia, Sudan, Syria, Tanzania, Yemen, Yugoslavia and Zambia.

[12] *See* the UN website, *http://untreaty.un.org/cod/repertory/art3/english/rep_supp5_vol1-art3_e.pdf* (visited on 01/06/2012).

[13] Further information about the A/L.630 is available at *http://www.un.org/en/ga/search/view_doc.asp?symbol=A/RES/2758(XXVI)&Lang=E&Area=RESOLUTION* (visited on 25/02/2014).

[14] *See* the UN website, *supra note* 12.

opposing resolution A/L.630 was submitted by 22 member states[15] on 29 September 1971. Resolution A/L.632 declared that any proposal in the General Assembly which would result in depriving the ROC of representation in the UN was an important question under Article 18 of the UN Charter, which would require it to be decided by a two-thirds majority of members present and voting.[16] On 25 October 1971, resolution A/L.632 was rejected.[17]

The United States was aware that the situation was favourable to the PRC, so through another section motion[18], it tried to retain the ROC's membership in the UN, resulting in two China representatives (implying two states) if resolution A/L.630 were adopted.[19] However, this motion failed, and the General Assembly adopted draft resolution A/L.630 by a roll-call vote of 76–35, with 17 abstentions, as resolution 2758 (XXVI)[20]. Before the adoption of the 2758 resolution, the ROC was aware that the resolution would be adopted so it decided to withdraw from UN membership in advance.[21] Since then, the ROC usually has been called "Taiwan" because the territories under its control are Taiwan and surrounding islands.

2.3 The Present—The Reality of International Relations

Although the ROC was compelled to withdraw from the UN, the competition between ROC and PRC to represent China in international

[15] These states are Australia, Bolivia, Colombia, Costa Rica, Dominican Republic, El Salvador, Fiji, Gambia, Guatemala, Haiti, Honduras, Japan, Lesotho, Liberia, Mauritius, New Zealand, Nicaragua, the Philippines, Swaziland, Thailand, the United States and Uruguay.

[16] *See* the UN website, *supra note* 12.

[17] *Ibid.*

[18] The representative of the United States subsequently proposed another motion that suggested that a separate vote be taken on including the words "and to expel forthwith the representatives of Chiang Kai-shek from the place which they unlawfully occupied at the United Nations and in all the organizations related to it" in the operative paragraph of draft resolution A/L.630.

[19] *See* Denny Roy, *supra note* 3, pp. 130–135.

[20] The full text of the A/RES/2758(XXVI) is available on the UN website, *http://daccess-dds-ny.un.org/doc/RESOLUTION/GEN/NR0/327/74/IMG/NR032774.pdf?OpenElement* (visited on 07/06/2012).

[21] *See* Peter S.C. Ho, "The Impact of the Fish Stocks Agreement on Taiwan's Participation in International Fisheries Fora," *Ocean Development and International Law*, Vol. 37(2006), p. 145, note 11.

forums has never stopped. The ROC's main strategy was to establish formal diplomatic relations with other states which recognise it as the legal representative of China.[22] However, since the PRC took the ROC's seat in the UN, the ROC's seat in other international organisations has also replaced by the PRC, and most states chose to transfer their formal diplomatic relations from the ROC to the PRC.[23] Therefore, the ROC gradually adopted another strategy of positively participating in international organisations, which it is easier and less costly to achieve than establishing diplomatic relations with states. The purpose of this strategy is to remind the international community that the ROC is an independent state not controlled by the PRC, to increase its visibility in international forum and to prevent being marginalised.

In contrast, to carry out the One China Policy, the PRC's strategy is to prevent the ROC from being recognised as a state. Accordingly, the PRC seeks to establish formal foreign relations with states that had foreign relations with ROC. The PRC also attempts to limit the ROC's international space, including rejecting any possibility for the ROC to participate in international organisations, especially those which might indirectly recognise the ROC as a state.

Facing opposition from the PRC, the ROC uses many different identities and names to join international organisations. For example, in

---

[22] One method the ROC uses to establish or maintain diplomatic relations with states is foreign aid. Sometimes, a state might exploit this weakness to request money from both the ROC and the PRC and then pick sides; this is called dollar diplomacy. The states which have established formal diplomatic relations with the ROC are Belize, Burkina Faso, the Dominican Republic, Republic of El Salvador, Republic of the Gambia, Republic of Guatemala, Republic of Haiti, The Holy See, Republic of Honduras, Republic of Kiribati, Republic of the Marshall Islands, Republic of Nauru, Republic of Nicaragua, Republic of Palau, Republic of Panama, Republic of Paraguay, Saint Christopher and Nevis, Saint Lucia, Saint Vincent and the Grenadines, the Democratic Republic of Sao Tome and Principe, Solomon Islands, Kingdom of Swaziland and Tuvalu.

[23] States that terminated diplomatic relations with the ROC usually changed their embassies to "offices" or "agencies" in order to still dealing with foreign affairs involving Taiwan. For example, the embassy of the ROC in the United States became the Taipei Economic and Cultural Representative Office in the United States, and the embassy of the United States in the ROC the American Institute in Taiwan, Taipei Office; the embassy of the ROC in Canada the Taipei Economic and Cultural Office in Canada, and the embassy of Canada in the ROC the Canadian Trade Office in Taipei; the embassy of the ROC in the UK the Taipei Representative Office in the UK, and the embassy of the UK in the ROC the British Trade and Cultural Office.

1991, the ROC used the name Chinese Taipei to join the APEC.[24] Then, it became a member of World Trade Organization (WTO) with the identity and name of the Separate Customs Territory of Taiwan, Penghu, Kinmen and Matsu on 1 January 2002.[25] It participated in the International Competition Network (ICN)[26] under the identity and the name of the Taiwan Fair Trade Commission in 2001. In 2010, the ROC joined the International Council for Information Technology in Government Administration (ICA)[27] under the identity of state named Taiwan.

In the maritime field, according to the Law on the Territorial Sea and the Contiguous Zone of the Republic of China[28], Taiwan claims that its territorial sea and contiguous zone extend 12 and 24 nautical miles from the baseline and outer limits, which is determined by a combination of

---

[24] The first occasion for the ROC to use the name "Chinese Taipei" to participate in an international organisation and avoid the political sensitivity of the One China Policy happened in 1981 when the ROC joined the International Olympic Committee (IOC), which had intended to block the ROC from continuing as a member. *See* Ho, *supra note* 21, pp. 145–146, note 17. All APEC members are called "economies," which identifies them as economic entities; hence, this organization does not raise the issue of the ROC's status as a state. *See* APEC, *http://www.apec.org/About-Us/About-APEC/Member-Economies.aspx* (visited on 08/06/2012).

[25] *See* WTO, *http://www.wto.org/english/thewto_e/countries_e/chinese_taipei_e.htm* (visited on 08/06/2012).

[26] *See* the ICN website, *http://www.internationalcompetitionnetwork.org/members/member-directory.aspx* (visited on 11/06/2012).

[27] *See* the ICA website, *http://www.ica-it.org/index.php?option=com_content&view=article&id=53&Itemid=85* (visited on 11/06/2012). Those are examples of international organisations that the ROC has joined. There are still many organisations in which Taiwan is a member, including the IATTC, Advisory Centre on WTO Law, Standards and Trade Development Facility, WCPFC, Extended Commission for CCSBT, ISC, Association of Asian Election Authorities, Asia/Pacific Group on Money Laundering, Central American Bank for Economic Integration, South East Asian Central Banks, Asian Development Bank, International Cotton Advisory Committee, International Seed Testing Association, Asian Productivity Organization, World Organization for Animal Health and International Organization of Securities Commissions.

[28] The Law on the Territorial Sea and the Contiguous Zone of the Republic of China was adopted and entered into force on 21 January 1998. The full text is available at *http://law.moj.gov.tw/Eng/LawClass/LawAll.aspx?PCode=A0000009* (visited on 02/04/2014).

straight baseline in principle and normal baseline as exception.[29] Under this law, the official maritime chart of the baseline and outer limits of territorial sea and contiguous zone of Taiwan was promulgated by the administrative department. In addition, pursuant to Article 2 of the Law on the Exclusive Economic Zone and the Continental Shelf of the Republic of China[30], Taiwan's EEZ is the sea area contiguous to the outer limits of the territorial sea and to a distance measuring outwardly 200 nautical miles from the baseline of the territorial sea. However, because Taiwan's EEZ overlaps with that of the PRC, the Philippines and Japan and there is no consensus on the borders of their EEZs, Taiwan has not promulgated a formal maritime chart of the outer limit of EEZ but in 2003 declared a temporary EEZ enforcement area in order to protect its fisheries within the EEZ.[31] The PRC, the Philippines and Japan did not publicly disagree with Taiwan's declaration concerning its territorial sea and temporary EEZ enforcement area. However, without consensus on delimitation of the overlapping EEZs, there remain many disputes among these states caused by fishing in these overlapping areas. According to official Taiwanese statistics, 2371 fishing vessels illegally trespassed in the temporary EEZ enforcement area in 2013; 97% of these vessels were Chinese nationals,

---

[29] *See* articles 3, 4 and 14 of the Law on the Territorial Sea and the Contiguous Zone of the Republic of China.

[30] The Law on the Exclusive Economic Zone and the Continental Shelf of the Republic of China was adopted and entered into force on 21 January 1998. The full text is available at *http://law.moj.gov.tw/Eng/LawClass/LawAll.aspx?PCode=A0000010* (visited on 02/04/2014).

[31] *See* the official website of the Coast Guard Administration, Executive Yuan, available at *http://www.cga.gov.tw/GipOpen/wSite/ct?xItem=5138&ctNode=891&mp=999* (visited on 02/04/2014). The temporary EEZ enforcement area was delimited pursuant to Article 4 of the Law on the Exclusive Economic Zone and the Continental Shelf of the Republic of China, which reads: "In the event that the exclusive economic zone or the continental shelf of the Republic of China overlaps with the adjacent or opposite countries, the Republic of China may negotiate, on the principle of equality, a delimitation line with those of the adjacent or opposite countries.

Prior to agreements mentioned in the preceding paragraph, the Republic of China and the adjacent or opposite countries, in a spirit of understanding and co-operation, may reach a modus vivendi.

Such a modus vivendi as prescribed in the preceding paragraph shall be without prejudice to the final delimitation."

and only 3% of other nationalities.[32] The enactment and declaration of a territorial sea, EEZ or even a temporary EEZ enforcement area reflect the reality of Taiwan's practice as a maritime actor. Although many Chinese vessels conducting illegal fishing within the Taiwanese temporary EEZ enforcement area are expelled, arrested and penalised, the PRC government does not respond to these events or attempt to oppose the enforcement by the Taiwanese government.

In 2013, Taiwan and Japan concluded a fishery agreement which does not delimit the EEZs but outlines an overlapping EEZ area between Taiwan and Japan and allows vessels of both nations to fish there.[33] The agreement was signed by the Association of East Asian Relations[34] and the Interchange Association,[35] quasi-official diplomatic units in Taiwan and Japan. This agreement not only illustrates again Taiwan's practice as a maritime actor but also reflects the reality that Japan *de jure* does not recognise Taiwan as a state but *de facto* recognises its position concerning

---

[32] In addition, 58% of these fishing vessels illegally trespassing were expelled by Taiwanese Coast Guard; the remaining 42% were arrested and penalised. Statistics concerning illegal trespass fishing can be found on the official website of the Coast Guard Administration, Executive Yuan, available at *http://www.cga.gov.tw/GipOpen/wSite/public/Attachment/f1394179647548.pdf* (visited on 04/04/2014).

[33] The title of this agreement is translated as The Fishery Agreement between the Association of East Asian Relations and the Interchange Association. The full text can be found on the official website of Ministry of Foreign Affairs of Taiwan, available at *http://no06.mofa.gov.tw/mofatreatys/ShowPicOut.aspx?FileFolder=00&FileName=002013/04/10_C.pdf* (visited on 04/04/2014).

[34] The Association of East Asian Relations, which was established in 1972, is an agency under the Ministry of Foreign Affairs of Taiwan. It deals with the foreign affairs between Taiwan and Japan. Further information can be found on the website of the Ministry of Foreign Affairs of Taiwan, available at *http://www.mofa.gov.tw/EnOfficial/Organization/DepartmentDetail/f510b84f-5e22-4fd5-b911-9d34141456b4* (visited on 08/04/2014).

[35] The Interchange Association was founded in 1972 when Japan established diplomatic relations with the PRC, which required terminating diplomatic relations with the ROC. The Interchange Association serves as the representative office of Japan in Taiwan, dealing with foreign affairs between Taiwan and Japan. The Interchange Association and the Association of East Asian Relations have functioned as the channel for negotiation and coordination between the two countries. Further information can be found on the Interchange Association's website, *http://www.koryu.or.jp/taipei-tw/ez3_contents.nsf/Top* (visited on 08/04/2014).

the EEZ and the temporary EEZ enforcement area; Japan thus can be regarded as treating Taiwan as a *de facto* state.

The methods employed to conduct Taiwan's relations with Japan can be observed in Taiwan's relations with many other states. The majority of states agree to follow the One China Policy and so *de jure* do not establish official diplomatic relation with Taiwan or recognise it as an independent state. However, they do not completely cut off ties with Taiwan but establish quasi-official diplomatic channels to deal with affairs with Taiwan. For example, states change the name of embassies to "offices" or "agencies." Such practices treat Taiwan as a *de facto* state.[36] Many international governmental organisations, including RFMOs, approach Taiwan similarly. Following the One China Policy, Taiwan is *de jure* refused participation in RFMOs but allowed to participate *de facto* under the identity of a fishing entity. This trend implies that although the international community follows the One China Policy, it acknowledges that Taiwan is not controlled by the PRC.

No matter what identity and name the ROC uses, it clearly participates in international organisations by emphasising its functions in different domains. Since the 1995 UNFSA introduced the term "fishing entities" into its provisions, Taiwan has seen that this term emphasises its function of fishing and may offer an opportunity to participate in RFMOs by using the identity of fishing entity. To understand Taiwan's practices in participating in RFMOs, including the process by which Taiwan accepts using the identity of fishing entity, the following section discusses relations between Taiwan and the 1995 agreement and Taiwan's participation in RFMOs with the identity of fishing entities with different names.

3. Relation between Taiwan and the United Nations Fishing Stocks Agreement

Although Article 1(3) of the UNFSA stipulates that it can apply to fishing entities whose vessels fish on the high seas, the agreement never defines the term "fishing entity." Neither is there any public record explaining to what the term "fishing entity" refers in the agreement.[37] However, the authors of some related literature either believe that the regulations

---

[36] For the examples, *see supra note* 23.

[37] *See* Nien-Tsu Alfred Hu, "Fishing Entities: Their Emergence, Evolution, and Practice from Taiwan's Perspective," *Ocean Development and International Law*, Vol. 37(2006), p. 153.

concerning fishing entities in the agreement were created especially for Taiwan or consider that they can be applied to Taiwan. Hasjim Djalal writes:

> It is clear that the provisions in UNFSA on fishing entities were intended primarily to deal with the fishing vessels flying the flag of Taiwan.[38]

David Anderson, leader of the UK delegation to the UNFSA conference, writes:

> The Agreement is seemingly intended to be applicable in principle to a "fishing entity" whose vessels fish on the high seas; Taiwan may have been in mind.[39]

Furthermore, while analysing the 1995 agreement, David A. Balton, legal adviser in the U.S. State Department, considers that

> Taiwan is not a member of the United Nations and did not participate in the conference except as an observer, despite the fact that many high seas fishing vessels are registered in its territory. Article 1(3) and 17(3) of the agreement provide a means to encourage compliance by vessels of Taiwan with rules of the agreement and with those of regional fishery organizations. Following adoption of the agreement, Taiwanese officials expressed willingness to abide by its parameters.[40]

Scholar Francisco Orrego Vicuña writes in his book:

> Another novel provision on participation is that applying the 1995 Agreement mutatis mutandis to "other fishing entities whose vessels fish on the high seas." While intended to cover the situation of Taiwan as a major fishing operator in the world ….[41]

He further comments:

---

[38] Hasjim Djalal, "The Emergence of the Concept of Fishing Entities: A Note," *Ocean Development and International Law*, Vol. 37(2006), p. 119.

[39] David H. Anderson, "The Straddling Stocks Agreement of 1995—An Initial Assessment," *International and Comparative Law Quarterly*, Vol. 45(1996), p. 468.

[40] David A. Balton, "Strengthening the Law of the Sea: The New Agreement on Straddling Fish Stocks and Highly Migratory Fish Stocks," *Ocean Development and International Law*, Vol. 27(1996), p. 149, footnote 73.

[41] Francisco Orrego Vicuña, *The Changing International Law of High Seas Fisheries* (Cambridge: Cambridge University Press, 1999), p. 139.

> The third step envisaged by the 1995 Agreement is to make its provisions applicable to other fishing entities. The 1995 Agreement applies mutatis mutandis to other fishing entities, a provision intended to take care of the particular legal situation of Taiwan.[42]

In the Letter of Submittal from the U.S. State Department to President W.J. Clinton on 24 January 1996, Warren Christopher, the secretary of state, points out that Articles 1(3) and 17(3) open a door for Taiwan to participate in RFMOs.

> Article 17(3), along with Article 1(3), also provides a mechanism through which Taiwan, and the many fishing vessels flying the Taiwanese flag, may be brought within the ambit of such organizations.[43]

These views indicate the link between the term "fishing entities" in the agreement and Taiwan. However, there is no public record further explaining fishing entities or the term's possible relation to Taiwan in the UNFSA Conference. The reason why cannot be stated with certainty but still needs to be traced to the political circumstances of Taiwan's status in international society. Nien-Tsu Alfred Hu, a member of the Presidential Advisory Council for Science and Technology in Taiwan, speculated that there is no public record and the term "fishing entities" is plural because "no one wanted to single out Taiwan at an international conference officially sponsored by and held at U.N. Headquarters." Although it is hard to prove whether this guess is correct, one thing which is certain is that the strong opposition of China to Taiwan creates its tough situation in the international community.

Nevertheless, although the UNFSA introduces the term "fishing entities" which can be applied by Taiwan, the Taiwan government initially worried whether the term would also cover private fishing firms or other nongovernmental organisations because Taiwan's core interest in participating in international organisations is to increase its visibility in international fora and present its case for independence from China.[44] These concerns were soon relieved because Article 1(3) shows that the agreement applies to only fishing entities and states. The provisions in Part

---

[42] *Ibid.*, pp. 212-213.
[43] *See* the Second Session of the 104th Congress in United States, Senate Treaty Document 104-24. The full text can be found at *http://www.gpo.gov/fdsys/pkg/CDOC-104tdoc24/pdf/CDOC-104tdoc24.pdf* (visited on 11/06/2012), p. X.
[44] *See* Hu, *supra note* 37, p. 155.

V on the Duties of the Flag State and Part VI on Compliance and Enforcement cannot be applied to private firms or nongovernmental organisations which lack the capacity to perform the responsibilities of a flag state or to cooperate with other states to conduct high seas fisheries enforcement.[45]

Furthermore, Article 17(3) indicates that fishing entities shall enjoy benefits from participation in the fishery commensurate with their commitment to comply with conservation and management measures adopted by RFMOs. From Taiwan's point of view, this article creates a fair condition because, if it is willing to fulfil obligations just like state parties, then it can also possess the same rights as they. This implies the possibility that fishing entities can have equal status as contracting parties in RFMOs, although Taiwan would soon realise that its status could never be equal because RFMOs do not regard fishing entities as states. They do, however, make the obligations and rights of a contracting party and a fishing entity as similar as possible.[46]

Before the adoption of the UNFSA, Taiwan was not a member of any RFMO. The 1995 agreement created an approach for Taiwan to participate in international fisheries organisations. Under the identity of fishing entity, Taiwan has become a member of the WCPFC, IATTC and ISC and has influenced other RFMOs of which it is not a member, such as the CCSBT and ICCAT.

4. Taiwan's Participation as a Member in Regional Fisheries Management Organisations

4.1 Western and Central Pacific Fisheries Commission

A. Taiwan's Participation

As the western and central Pacific lacked a RFMO to manage fishery resources, the Pacific Islands Forum Fisheries Agency[47] called for a conference on south Pacific tuna fisheries to adopt a convention and establish a RFMO. This call led to the seven sessions of the Multilateral High-Level Conference in the Conservation and Management of Highly Migratory Fish Stocks in the Western and Central Pacific (MHLC) from 1994 to 2000. The process of drafting the WCPFC Convention concerning

[45] *Ibid.*

[46] *Ibid.* Hu suggests that Article 17(3) provides a balanced approach to solving the problem for Taiwan as a non-contracting party.

[47] *See* Chapter 3, fn 2, of this book.

Taiwan's participation in the MHLCs reflected the intense competition between the ROC and the PRC. The first MHLC was held in December 1994 in the Solomon Islands, a state which had formal diplomatic relations with the ROC and recognised it as a state. Taiwan saw it as a good opportunity to participate under its official name, the ROC.[48] However, this was not allowed because it conflicted with the One China Policy adopted by the PRC and the states with diplomatic relations with it. As a result, Taiwan refused to participate in this meeting.[49]

Taiwan started participating in the MHLC2, held in 1997 in the Marshall Islands which also had official relations with the ROC. During the process of negotiating Taiwan's participation, the PRC suggested the names of Taiwan, China or China-Taiwan, but Taiwan refused both and eventually agreed to use the name Chinese Taipei which was also accepted by most participants.[50] Interestingly, the delegation of Chinese Taipei was seated at the table under the letter "T," between Solomon Islands and Tonga, in order to not only avoid the embarrassment of being seated next to the PRC delegation but also to implicitly reflect Taiwan's actual status as not subordinate to China but equal to other participating states.[51] As another compromise, national flags were not displayed inside or outside the meeting room, nor were official designations listing participants in the meeting recording.[52]

In its opening statement at MHLC2, Taiwan stated that it would participate under the identity of a fishing entity:

> As a fishing entity defined in the [UNFSA] and as an active fishing nation[53] in the region, we see ourselves as a constructive force in the formulation of any possible regional arrangement which satisfies the mutual interests of both coastal and fishing nations. We intend to collaborate with all parties concerned on a basis of equality, full participation and full membership to work out a regional arrangement

---

[48] Michael W. Lodge, "The Practice of Fishing Entities in Regional Fisheries Management Organizations: The Case of the Commission for the Conservation and Management of Highly Migratory Fish Stocks in the Western and Central Pacific Ocean," *Ocean Development and International Law*, Vol. 37(2006), p. 187.

[49] *Ibid.*

[50] *Ibid.* Also *see* Hu, *supra note* 37, p. 158.

[51] Lodge, *supra note* 48, p. 188.

[52] *Ibid.*

[53] Here, "fishing nation" means a fishing state. This designation reflects Taiwan's position that it regards itself as a state, although it uses the identity of fishing entity to participate in the meeting.

> which reflects and embodies the spirit and letter of [UNCLOS] and the [UNFSA].[54]

Since Taiwan was the only participant with the identity of fishing entity, it is clear that, at least in the following MHLCs, the discussion or articles concerning fishing entities specifically referred to Taiwan. The other significant matter regarding Taiwan in the MHLC2 was its adoption of the Majuro Declaration, which put fishing entities in the same position as states[55] and implied that fishing entities possess rights and obligations equal to coastal and other states under the UNFSA.

In an opening statement, the chairman expressed the necessity of Taiwan's participation to increase the effectiveness of fisheries conservation and management of the WCPFC and further called on all participants to restrain from raising political issues.

> If we are going to have an effective regime for fisheries conservation and management in the region then it is obvious that all those who belong to the region or fish in the region must be involved. The participation of Chinese Taipei in this regard is on that basis. The present arrangement for participation, however, does not determine its final relationship to any agreement that this Conference might adopt. In arriving at such a relationship I shall certainly consult with all parties concerned in order to find a solution which is not only practical and realistic but also appropriate. In the meantime I urge all participants not to raise political issues which detract us from the basic purpose of our meeting here which is to agree on a regime for the conservation and management of the highly migratory species in the central and western Pacific. I therefore ask for restraint on all sides in this matter.[56]

However, when the chairman's proposed Draft Convention on the Conservation and Management of Highly Migratory Fish Stocks in the Western and Central Pacific Ocean began to be discussed during MHLC4 in February 1999, the PRC exercised its political influence to limit Taiwan to participating as an observer under the designation of Taipei, China instead of Chinese Taipei.[57]

At the end of the MHLC4, the chairman proposed a "Draft Preamble and Final Clauses for a Convention on the Conservation and Management of Highly Migratory Fish Stocks in the Western and Central Pacific." It

---

[54] Hu, *supra note* 37, p. 158.
[55] Regarding the Majuro Declaration, *see* Chapter 3 of this book.
[56] Hu, *supra note* 37, p. 159.
[57] *Ibid.*

would allow fishing entities, after the convention has entered into force, to accept the regime of the convention and open the commission to full participation by those fishing entities. Article 43(2) reads: "After the entry into force of this Convention, the Commission shall also be open to full participation by fishing entities referred to in Article 35, paragraph 3, which have affirmed their acceptance of the regime established by this Convention, in the same manner as members of the Commission."[58] This draft of the final clause was discussed at MHLC5 in September 1999 but was rejected by both Taiwan and the PRC.

In the draft final clause at MHLC4, the PRC could not accept the provision allowing fishing entities to affirm their acceptance of the convention under the title of "Signature, ratification, formal confirmation, acceptance, approval, accession" because such formal treaty language might imply that Taiwan, as a possible fishing entity, could be regarded as possessing equal status with states.[59] On the other hand, Taiwan was not satisfied with the notion of "full participation," which was also used in territories' participation, in Article 43(2). Taiwan insisted on becoming a full member of the commission with the same rights and obligations as contracting parties.[60] Taiwan could not be confident whether it could obtain full membership status in the commission through "full participation."[61] Contradictorily, the PRC could not accept the notion of "full participation" because it was too far from the status of an observer on which it insisted for Taiwan.[62] In addition, the draft granted Taiwan participation through expressing its consent to the "regime established by the Convention" rather than the convention itself; thus, Taiwan doubted whether it had liberty to withdraw from the regime once it had accepted it.[63]

---

[58] *Ibid.*, p. 160.
[59] Lodge, *supra note* 48, p. 191.
[60] *Ibid.*
[61] Hu, *supra note* 37, p. 160.
[62] Lodge, *supra note* 48, p. 191.
[63] Therefore, Taiwan expressed its strong negative reaction to the draft in the opening statement:

> *The texts of the draft Convention....with respect to the participation status of any fishing entity to the Convention are conceivably not workable for any democratic government. The rights and obligations of a fishing entity under the present framework are not parallel, and the requirements for obligations and compliance ensued from the present draft Convention will definitely have relevance to the assignment of sovereignty and sovereign rights to an international fisheries management body. On these two grounds, the executive and legislative bodies of a democratic government*

To solve this problem, the chairman at MHLC5 issued a Revised Draft Convention on the Conservation and Management of Highly Migratory Fish Stocks in the Western and Central Pacific Ocean, which retained the provisions of Article 35(3) but moved it to Article 42 under the title "Fishing Entities," instead of "Signature, ratification, formal confirmation, acceptance, approval, accession." Furthermore, this revised draft convention distinguished between contracting parties and member of the commission in Article 1(1)(d): "'Contracting Parties' means States and regional economic integration organizations which have consented to be bound by this Convention and for which the Convention is in force."[64] Article 1(2)(b) reads: "'Member of the Commission' means a Contracting Party and, as appropriate, any fishing entity referred to in article 42, which has affirmed its acceptance of the regime established by this Convention. Membership of the Commission does not in any way determine the legal or political status of any such fishing entity."[65] However, sensitive issues concerning Taiwan's participation were still not resolved. With respect to Taiwan's concern about withdrawal of its consent to the regime established by the convention, Article 41 of the Revised Draft Convention stipulated that a "member of the Commission may, by written notification addressed to the Depositary, withdraw from this Convention and may indicate its reasons."[66]

Before MHLC6, the chairman issued an Information Note on Matters before the Sixth Session of the MHLC describing how a fishing entity could become a member of the commission and possess the same rights and obligations as a contracting party:

> A significant amount of time during the fifth session was devoted to the issue of participation in the work of the Commission by fishing entities.

---

will not be able to endorse the draft Convention as the way it is. This situation will unfortunately force out a fishing entity with a democratic institution from the present Convention—a situation that no one likes to embrace.

Mr. Chairman, we believe that the design and the languages of the present draft Convention with respect to the participation of a fishing entity to the Convention is not workable for us, which, in turn, may damage the future work and operation of the Convention itself. See Hu, *supra note* 37, p. 160.

[64] MHLC/WP.1/Rev.4, available at *http://www.spc.int/coastfish/Asides/conventions/MHLC/mhlc_sep99.htm* (visited on 01/11/2010).

[65] *Ibid.*

[66] Hu, *supra note* 37, p. 161.

> The draft Convention text would allow a "fishing entity, being a separate customs territory possessing full autonomy in the conduct of its external commercial relations, and whose vessels fish for highly migratory fish stocks in the Convention Area" to become a full member of the Commission with the same rights and obligations as a Contracting Party after affirming in writing its acceptance of the Convention regime. Such affirmation may be made only after entry into force of the Convention. The draft Convention text further provides that "[M]embership of the Commission does not in any way determine the legal or political status of any such fishing entity." While the exchange of views on this issue was constructive, this sensitive issue remains to be concluded.[67]

The conflict between the ROC and the PRC reached a climax at MHLC6 in April 2000. In an opening statement, Taiwan stated:

> [W]e have long actively and constructively participated in the activities of various regional and/or subregional fisheries organizations since we, as a major global fishing power, recognize the importance of conservation of fisheries resources of the high seas to all the fishing nations, and take into account the cooperative needs required by the international community as a whole. We believe that our equal participation and substantive cooperation could have at a very large extent contributed to the effectiveness of conservation and management of fisheries resources concerned. … We would like to highlight the value of conservation and management of this particular "common property," the fisheries resources in the seas and oceans, and to contribute more to the realization of this value. …
>
> Mr. Chairman and my colleagues, I understand that you all are watching with great interest the recent development in Taiwan, especially after the presidential election. The result of the election once again proves to the world that a genuine democratic institution is functioning in Taiwan. At the same time, we anticipate that both sides of the Taiwan Straits will enter the World Trade Organization as two full members in this year. Against the backdrop of this new development, we believe that a Contracting Party status to the Convention for us is needed in order to secure congressional approval in our democratic institution.
>
> In considering the issue of our participation status, we have to recognize the fact that we are the one fully capable of exercising control over our own fishing vessels and are competent of exercising jurisdiction over fisheries matters provided for in this draft Convention. Thus, under the

[67] *See* paragraph 18 of the "Information Note on Matters before the Sixth Session of the Multilateral High-Level Conference on the Conservation and Management of Highly Migratory Fish Stocks in the Western and Central Pacific," in the Report of MHLC6, p. 37.

> framework of this Convention, we are competent to bear the obligations and are entitled to enjoy the rights as other parties to this Convention. Mr. Chairman and dear colleagues, these are the crucial matters in considering our position vis-à-vis this regional regime or organization. The needs for a comprehensive and inclusive fisheries conservation and management mechanism has been over and over again underscored by our Chairman Ambassador Nandan during the development of the 1995 [UNFSA]. Indeed, to constitute a comprehensive and inclusive instrument with regard to and with intention for an effective conservation and management of highly migratory fish stocks in this region, it would be a major defect were Taiwan not to be rendered a Contracting Party status, should this Conference so agree. It would certainly undermine the purpose and the effectiveness of this regional scheme that we are endeavoring to formulate.[68]

The PRC responded:

> Mr. Chairman, this morning, my colleague from Chinese Taipei mentioned the election in the island. We hold that the local election in Taiwan and the element of democracy and other irrelevant matters have nothing to do with fish stock conservation and management; and cannot be served as a reason for Chinese Taipei to participate in the future Commission, the composition of which should be based on UNCLOS and [UNFSA]. As a matter of fact, our colleague from Chinese Taipei has politicized the fishing issues in his opening statement on purpose. We are surprised and feel uncomfortable about it. We hold that our colleague from Chinese Taipei should honour the rule of the game that has been agreed upon by all participants, including Chinese Taipei, from the first session of the MHLC. Such attitude is by no means constructive to us. In order to spare more time to discuss other significant pending issues, we hope the Conference shall not allow this phenomenon to reoccur.
>
> The Chinese Government is of the opinion that fishing entities should be made to comply with the conservation regulations in the Convention area. But this should not impair the sovereignty of the State concerned, or it may lead to numerous political troubles in the future implementation of the Convention. We hold that the status of fishing entities has no inevitable and direct link with the conservation and management measures. We wish the Convention be worked out in the guidance of the principles of UNCLOS

[68] *See* the Opening Statement of Chinese Taipei, in the Report of MHLC6, pp. 27-28. Also s*ee* Andrew Serdy, "Bringing Taiwan into the International Fisheries Fold: the Legal Personality of a Fishing Entity," in James Crawford and Vaughan Lowe, eds., *The British Year Book of International Law 2004* (Oxford: Clarendon Press, 2005, Vol. 75), pp. 210–213.

> and [UNFSA], and the spirit of consensus as well to ensure a smooth adoption.[69]

The conflict and contradiction between Taiwan and the PRC was reflected not only in their statements but also in those by other states' opening statements. Nine states with diplomatic relationships with or friendly to Taiwan either supported Taiwan in obtaining contracting party status or in participating fully. The other twelve delegations took no position.[70] During MHLC6, Taiwan expressed its anticipation of holding the same position, including the rights and obligations in the commission, as a contracting party, as expressed in two documents entitled Chinese Taipei's Views on Our Participation as a Contracting Party to the MHLC Convention and Legal Arguments for the Participation of a Fishing Entity in Regional Fisheries Management Organisations in the Capacity of Contracting Party.[71] In contrast, the PRC proposed Our Position on the Position of Fishing Entity and a specific article concerning fishing entities. Both suggested that Taiwan become an observer of the commission and tried to prevent Taiwan from gaining the same status as other independent states. The documents would require the following text to be inserted in the draft convention:

> 1. After the entry into force of this Convention, any fishing entity, whose vessels fish for highly migratory fish stocks in the Convention Area, may confirm in writing to cooperate fully in the implementation of conservation and management measures adopted by the Commission. Such fishing entities shall enjoy benefits from the fishery commensurate with their commitment to comply with

---

[69] *See* the Opening Statement of China, in the Report of MHLC6, pp. 11–12. Also *see* Serdy, *supra note* 68, pp. 211–212. According to the record of Hu, a member of the Taiwanese delegation, the PRC requested the floor again using emotional language that did not appear in the official record, including the statements: "Adolf Hitler was also elected by popular votes," "Do not squeeze your Central Government" and "Barrels of gun powder are ready for a small handful of separatists, if they dare to try it." *See* Hu, *supra note* 37, p. 162.

[70] Five states Nauru, Palau, Solomon Islands, Tuvalu and the United States, supporting Taiwan in becoming a contracting party, and four states, Fiji, Marshall Islands, Papua New Guinea and Vanuatu, supported Taiwan's full participation. The other 12 delegations were from Canada, the Federated States of Micronesia, Indonesia, Japan, Kiribati, Republic of Korea, New Zealand, the Philippines, Mexico, French Polynesia, New Caledonia and the European Community.

[71] Lodge, *supra note* 48, p. 192.

conservation and management measures in respect of the relevant stocks.

2. Fishing entities referred to in paragraph 1, may upon request and subject to the concurrence of the members of the Commission and to the rules of procedure relating to the granting of observer status, be invited to attend meetings of the Commission as observers.[72]

The views of Taiwan and the PRC could not be reconciled, making for difficult concerning the position of fishing entities difficult. The chairman consulted separately with both sides but still failed to find a compromise.[73] The difficult situation is exhibited in the chairman's information note before MHLC7.

> The question of the relationship of fishing entities to the draft Convention has always been a particularly difficult issue. Unfortunately, it was not possible to make progress on this issue during the sixth session. The issue is a sensitive political matter for the two parties concerned, but it is equally important for all other participants in the Conference because they want to ensure that in any conservation and management regime all major actors are involved and comply with the regime that is being established. Most participants in the Conference would like to see an outcome which creates legal obligations on fishing entities and enables substantive participation. The difficulty is how to find a formula which creates a legally binding relationship without prejudging the legal and political status of such entities. At the end of the sixth session the Chairman had requested all concerned to give further thought to this matter and to show some flexibility.[74]

Another note from the chairman again states that the matter could go further; at a meeting with Japan, the United States, the PRC and Taiwan, the Chairman indicated that:

> Neither Japan nor the US had any new suggestions on this issue. Both feel that the chairman's current proposal provides a basis for compromise between China and Chinese Taipei. I wish also to inform you that I have discussed the matter with both China and Chinese Taipei in recent weeks with the intention of clarifying their respective positions. While I have no

[72] *Ibid.*

[73] *Ibid.*, p. 193.

[74] *See* paragraph 17 of the "Information Note on Matters before the Seventh Session of the Multilateral High-Level Conference on the Conservation and Management of Highly Migratory Fish Stocks in the Western and Central Pacific," in Annex 4 of the Report of MHLC7.

> further solutions to offer at this time, I will continue to work on the issue prior to MHLC7 and have urged both China and Chinese Taipei to do likewise.[75]

In the opening statement of MHLC7 in August 2000, the PRC reiterated its position that a fishing entity could become only an observer.

> [M]y delegation cannot accept the formulation on the issue of fishing entities in the current text of the draft Convention. The issue of fishing entities has been originated from the 1995 [UNFSA]. In accordance with [UNFSA], the issue of fishing entities should be deferred until the Commission is established for its consideration. In view that my delegation has agreed to invite the fishing entity to participate in the MHLC, my delegation is prepared to join the effort in making appropriate arrangement for fishing entities to participate in the future regime of the Convention for the purpose of both ensuring fishing entities bound by the conservation and management measures of the Convention and safeguarding the interests of fishing entities. However, my delegation cannot accept the view that the issue of fishing entities is a bilateral political issue; my delegation cannot accept the attempt to raise the political status of fishing entities; and my delegation cannot accept the move of some people to use the issue of fishing entities for other interests. Fishing is a commercial business. No matter how large the fishing fleet is, fishing is an industry. Fishing entities' interests is fishing. ... My delegation firmly believes that, as the Convention is intended to provide for the conservation and management measures for highly migratory fish stocks in western and central Pacific, all the fishing fleets should be bound by these measures but the Convention should not become an opportunity for fishing entities to raise status. Fishing entities could be involved in the work of the Commission but can do so only in the capacity of observer.[76]

Subsequently, the chairman held several informal and bilateral private meetings, but the ROC and the PRC never met face to face.[77] During the negotiation process, Taiwan felt that it had already made important political and legal compromises in its agreement to participate under the identity of a fishing entity and the name Chinese Taipei as its official nomenclature.[78]

---

[75] Hu, *supra note* 37, p. 164.
[76] The full text of statement is available at *http://www.spc.int/coastfish/Asides/Conventions/MHLC/MHLC7rep.pdf* (visited on 09/11/2010), p. 14.
[77] Hu, *supra note* 37, p. 165.
[78] Lodge, *supra note* 48, p. 193.

The chairman indicated that most other participants did not accept the PRC's insistence on observer status for fishing entities.[79] With respect to the decision-making matters, the PRC argued that Taiwan could not help decide the election of chairman and vice-chairman, location of headquarters, appointment of executive director and adoption of rules, financial regulations and a budget.[80] Taiwan successfully argued that it was unfair to exclude it from deciding the budget and financial regulations because it would be a major contributor to the budget of the commission. The PRC agreed to comprise that Taiwan could participate in the election of a chairman, provided that the chairman was selected from the contracting parties.[81] In addition, while considering whether the convention should list all the matters in which Taiwan could participate in decision-making or the matters in which Taiwan could not, the Chairman proposed the latter.[82] As a result, the matters in which Taiwan could not participate were only the location of the headquarters and the appointment of the executive director.

Regarding the form and the nomenclature of the instrument that Taiwan could sign, the chairman preferred to use "declaration" instead of "protocol," which the PRC viewed with concern because it had connotations of a treaty.[83] Taiwan could not accept "declaration" because of its unilateral connotation. Instead of "declaration," Taiwan suggested "Arrangement to Facilitate the Participation of Fishing Entities," which the chairman immediately accepted with slight modification as "Arrangement for the Participation of Fishing Entities."[84]

During discussion of the remaining problems, it was agreed to put the detailed provisions concerning fishing entities into an annex to the convention.[85] Subsequently, a revised text of the Draft Convention on the Conservation and Management of Highly Migratory Fish Stocks in the Western and Central Pacific Ocean was proposed and adopted on 4 September 2000. Taiwan signed the arrangement as a fishing entity by the name of "Chinese Taipei" on 5 September 2000 and became a member when the WCPFC was established on 19 June 2004.

---

[79] *Ibid.*
[80] *Ibid.*, p. 196.
[81] *Ibid.*, pp. 196-197.
[82] *Ibid.*, p. 196.
[83] Hu, *supra note* 37, p. 165.
[84] *Ibid.*
[85] Lodge, *supra note* 48, p. 196.

B. High Seas Boarding and Inspection

Pursuant to Article 26 of the WCPFC Convention, the WCPFC Boarding and Inspection Procedure was adopted in 2006. In the procedure, the subjects which can be inferred to have the rights to board and inspect other members' vessels are defined by the term "contracting party," and the subjects which can be inferred to have obligations to accept being boarded and inspected are described by the term "member of the commission." As a result, contracting parties of the WCPFC Convention and the members of the commission might have different legal status under the WCPFC Boarding and Inspection Procedure.

During the drafting of the Procedure, it was not decided whether the subject of the procedure was "member of the commission" or "contracting parties." Taiwan suggested not adopting the latter. In a letter to the Convener of the Intersessional Working Group on High Seas Boarding and Inspection Procedures, Mr. William Gibbons-Fly, Taiwan expressed its position that the authority to board and inspect fishing vessels on the high seas in the convention area should be available to all members of the commission.[86] It mentioned:

> [T]he restriction of the competence to board and inspect fishing vessels on the high seas in the Convention Area to Contracting Parties alone was not sustainable and was subsequently defeated, and thus not appeared in the final draft for the adoption. … It should not be argued again that the right to boarding and inspection is to be limited to Contracting Parties alone if the WCPFC Convention is going to be interpreted in good faith.[87]

However, the final version of the Boarding and Inspection Procedure was adopted as it still stands. It can be assumed that Taiwan objected to this outcome, but there is no record of how the decision was made or of the turning point from favouring member of the commission to contracting parties.

Pursuant to Article 6 of the Procedure and Paragraph 8 of the Revised Procedure in 2007, Taiwan and a contracting party which exchange

---

[86] *See* WCPFC-TCC2-2006-12, High Seas Boarding and Inspection Procedures (Attachment 2-Chinese Taipei Comments 16 March 2006) and WCPFC-TCC2-2006-12, High Seas Boarding and Inspection Procedures (Attachment 2–Chinese Taipei Comments 17 March 2006), available at *http://www.wcpfc.int/doc/wcpfc-tcc2-2006-12/high-seas-boarding-and-inspection-procedures-attachments* (visited on 15/08/2012).

[87] WCPFC-TCC2-2006-12, High Seas Boarding and Inspection Procedures (Attachment 2–Chinese Taipei Comments 17 March 2006), p. 4.

notifications may board and inspect each other's vessels.[88] Nevertheless, it is unclear whether the vessels of Taiwan can be boarded and inspected by a contracting party which does not make a notification, in other words, whether Taiwan needs to comply with the WCPFC Boarding and Inspection Procedure if there is no notification. For Taiwan, Article III of the Regulations for Coast Guard Administration to Conduct WCPFC High Seas Boarding and Inspection Procedures, [89] dated 1 May 2010, corresponds to the related regulations of WCPFC but fails to further explain them. This articles states that "[t]he boarding and inspection activities performed under the regulations are only applicable to fishing vessels whose countries agree to apply the WCPFC boarding and inspection procedures on high seas with Republic of China." The answer is not revealed in the related regulations of either the WCPFC or Taiwan.

In practice, Taiwan has 13 registered inspection vessels. The first nine were circulated by the executive director of WCPFC on 15 May 2008 and qualified to start boarding and inspection on 14 July 2008. Two other vessels were circulated on 1 June 2010 and on 23 March 2011 respectively, and qualified to start boarding and inspection on 1 August 2010 and 21 May 2011. The last two were circulated on 3 April 2013 and qualified to start boarding and inspection on 1 June 2013 (See Table 7-1).[90] With respect to notification from the contracting parties, by June 2012 six states—New Zealand (1 August 2008), Cook Islands (21 November 2008), the United States (12 August 2009), Japan (7 October 2009), France (4 December 2009) and Australia (on 29 September 2010)—had sent notification letters to the commission, expressing their consent to apply the WCPFC Boarding and Inspection Procedure between Taiwan and them.[91] The 1–6 October 2009 Report of the Fifth Session of the Technical and Compliance Committee (TCC5) mentions the conduct of boarding and inspection by Taiwan. It states that "Chinese Taipei deployed an inspection

[88] For an analysis of Article 6 of the Procedure and Paragraph 8 of the Revised Procedure in 2007, please *see* Chapter 6 of this book.

[89] The full text can be found on the official website of the Coast Guard Administration in Taiwan, *http://www.cga.gov.tw/wralawgip/cp.jsp?displayLaw=true&lawId=8a8181d92b5668a7012b5c2785200005&printPage=true* (visited on 06/07/2012).

[90] Details about these vessels, including their registration number, port of registry, vessel length, hull material, contact phone number and photographs, can be found on the official WCPFC website, *http://www.wcpfc.int/high-seas-boarding-inspection* (visited on 27/05/2013).

[91] These notifications are available at the official WCPFC website, *http://www.wcpfc.int/high-seas-boarding-inspection* (visited on 27/05/2013).

vessel for 89 days beginning in 6 October 2008 and has deployed two inspection vessels in 2009. These vessels have focused on HSBI [High Seas Boarding and inspection] of Chinese Taipei and other vessels in the high seas pockets and on instructing the vessels to comply with WCPFC CMMs [Conservation and Management Measures]."[92]

Table 7-1

	Name of Taiwanese Inspection Vessel	**Circulation Date by Executive Director**	**Starting Date of Boarding and Inspection Pursuant to CMM 2006–08**
1	SHUN HU NO.1	15 May 2008	14 July 2008
2	SHUN HU NO.2	15 May 2008	14 July 2008
3	SHUN HU NO.3	15 May 2008	14 July 2008
4	HO HSING	15 May 2008	14 July 2008
5	WEI HSING	15 May 2008	14 July 2008
6	TAI PEI	15 May 2008	14 July 2008
7	NAN TO	15 May 2008	14 July 2008
8	KIN MEN	15 May 2008	14 July 2008
9	LIEN CHIANG	15 May 2008	14 July 2008
10	YU SHIUN NO.2	1 June 2010	1 August 2010
11	HSUN HU NO.7	23 March 2011	21 May 2011
12	HSUN HU NO.8	3 April 2013	1 June 2013
13	HSUN HU NO.9	3 April 2013	1 June 2013

Source: WCPFC website, *http://www.wcpfc.int/high-seas-boarding-inspection* (visited on 27/05/2013).

In 2010, two inspection vessels of Taiwan boarded and inspected two US-flagged vessels and 16 Chinese Taipei-flagged vessels, but no violations were observed.[93] In 2011, Taiwan deployed three vessels for a

[92] See the Summary Report of the Fifth Session of the Technical and Compliance Committee, p. 30, available at *http://www.wcpfc.int/meetings/2009/5th-regular-session-technical-and-compliance-committee* (visited on 05/07/2012).

[93] *See* the Summary Report of the Sixth Session of the Technical and Compliance Committee, p. 30, available at *http://www.wcpfc.int/meetings/2010/6th-regular-session-technical-and-compliance-committee* (visited on 05/07/2012).

total of 270-day patrols and conducted boarding and inspection of 44 Chinese Taipei-flagged vessels, but no violations were found.[94]

It is worth noting that, in the 7th Meeting of the WCPFC, the PRC sent a letter to the WCPFC executive director, objecting to the viewpoint of Taiwan and a contracting party[95] that reciprocally implemented the WCPFC Boarding and Inspection Procedure.

> Dear Mr. Chairman:
>
> ……….
>
> China noted that a Contracting Party stated in its notification to the Commission that "absent a notification by a Contracting Party of its intent to apply the Procedures *on a reciprocal basis* as between the Contracting Party and a Fishing Entity, the Procedures are not in effect as between the Contracting Party and the Fishing Entity." (emphasis added)
>
> China also noted that a Fishing Entity echoed that above-mention viewpoint. In its response to some Contracting Parties' notification under paragraph 6 of the Procedures, the Fishing Entity emphasized that in the absent a notification by a Contracting Party of its intent to apply the Procedures in its entirety as between that Contracting Party and a Fishing Entity on a reciprocal basis, the Procedures are not in effect as between the Contracting Party and the Fishing Entity.
>
> China wishes to point out that nothing in the Procedures provides that the high seas boarding and inspection activities between a Contracting Party and a Fishing Entity are conducted on a reciprocal basis. China does not accept the viewpoint of the Contracting Party and the Fishing Entity in this regard.[96]

It is clear that the PRC would not like to see the WCPFC Boarding and Inspection Procedure applied between Taiwan and a contracting party only if the contracting party gave notification. In other words, the PRC thought that, if a contracting party exchanged notification under Article 6, then this contracting party and Taiwan could board and inspect each other's vessels. However, if a contracting party did not make a notification, then it could

---

[94] *See* the Summary Report of the Seventh Session of the Technical and Compliance Committee, p. 25, available at *http://www.wcpfc.int/doc/TCC7-Summary-Report-%28Rev-1%29* (visited on 28/05/2013).

[95] The PRC did not name the state, but from its description comparing the notifications of those six states, it can be inferred to refer to the United States. The notification of the United States can be found at *http://www.wcpfc.int/high-seas-boarding-inspection* (visited on 06/07/2012).

[96] *See* WCPFC-2010-IP-08, available at *http://www.wcpfc.int/doc/wcpfc7-2010-ip-08/chinas-letter-wcpfc-executive-director* (visited on 06/07/2012).

still board and inspect Taiwanese vessels, but Taiwan would not have the right to do the same to that contracting party.

Interestingly, Taiwan's statement responding to the PRC's letter did not directly touch the point which concerned the PRC.

> The WCPFC Boarding and Inspection Procedures (hereafter CMM 2006-08) were adopted in accordance with Article 26 of the WCPFC Convention. These procedures set out the WCPFC's "boarding and inspection scheme" and are supposed to "effectively discharge the obligations of the members of the Commission under the Agreement and this Convention to ensure compliance with the conservation and management measures established by the Commission." These provisions, as contained in CMM 2006-08, have been very successful on avoiding political issues.
>
> According to paragraph 6 of the CMM 2006-08, a Contracting Party of the WCPFC can decide whether or not to apply these procedures "in their entirety" with respect to a Fishing Entity. Thus far, six Contracting Parties have taken this approach, which ensures full compliance with the conservation and management measures established by the Commission.[97]

Taiwan avoided causing a dispute with the PRC which could reopen the political issue without favouring Taiwan given the PRC's political power and influences. Furthermore, once this issue were handled in the way the PRC suggested, it would be disadvantageous to Taiwan. Therefore, the best strategy for Taiwan is to maintain the ambiguous state of affairs so that it can have greater flexibility in dealing with interpretations of the related provisions.

4.2 Inter-American Tropical Tuna Commission

During 15–16 October 1998, the IATTC adopted the resolution The Participation of Taiwan in the work of the Inter-American Tropical Tuna Commission,[98] welcoming Taiwan's commitment to participate actively in the work of the IATTC and recommending that member governments find the most appropriate mechanisms to enable Taiwan's active participation. Hence, Taiwan was invited to participate as a negotiating partner on equal

---

[97] *See* WCPFC-2010-IP-16. The full text is available at *http://www.wcpfc.int/doc/wcpfc7-2010-ip-16/chinese-taipeis-letter-wcpfc-chairman* (visited on 06/07/2012).

[98] *See* the Resolution 98-09, available at *http://www.ofdc.org.tw/fishserv/File/Rule/IATTC_Resolutions_E/98-09.pdf* (visited on 15/06/2012).

footing with other IATTC convention parties in the meetings of the *ad hoc* working group, established in 1998 with the aim to modify the 1949 IATTC Convention.[99] For Taiwan, the moment offered an advantageous opportunity to seek status as a contracting party in the IATTC because five existing contracting parties or members had diplomatic relations with Taiwan, and US policy supported Taiwan in joining international organisations with a functional purpose, such as RFMOs.[100] In January 1999, at the second session of the working group, the question of Taiwan's participation was first raised. The Nicaraguan delegate proposed that the term "parties" should encompass three categories: states, regional economic integration organisations and separate customs territories.[101] Taiwan found this proposal acceptable because it could participate in the new IATTC Convention as a party under the identity of a separate customs territory, following the model from the WTO Agreement[102]. Taiwan preferred the WTO model because it enjoyed full membership and equal status with other states in the WTO and hoped to replicate this formula in the IATTC.[103] Moreover, the WTO model had been accepted by international forums and in domestic legislation and so could set a precedent that might make it easier to overcome political difficulties, both internationally and locally.[104]

However, the Ecuadorian delegate noted that the term "separate customs territories" was from the WTO, which was designed to deal with trade issues, and questioned whether it was appropriate for used in a fishery resources management organisation.[105] In response, the Taiwan delegate argued that the international trade of fish and fish products were also an important link in the fishery and cited the regulations of the 1995

---

[99] About the establishment of the Working Group, *see* the Resolution 98-02, available at *http://www.ofdc.org.tw/fishserv/File/Rule/IATTC_Resolutions_E/98-02.pdf* (visited on 15/06/2012). On Taiwan's participation. *see* Hu, *supra note* 37, p. 167; Dustin Kuan-Hsiung Wang, "Taiwan's Participation in Regional Fisheries Management Organizations and the Conceptual Revolution on Fishing Entity: The Case of the IATTC," *Ocean Development and International Law*, Vol. 37(2006), p. 210.

[100] Ho, *supra note* 21, pp. 142-143.

[101] This proposal was supported by Costa Rica and El Salvador, whereas Venezuela proposed that "parties" should include only states and regional economic integration organisations. *See* Wang, *supra note* 99, p. 210.

[102] The full text of the WTO Agreement can be found on the official WTO, *http://www.wto.org/english/docs_e/legal_e/04-wto.pdf* (visited on 16/08/2012).

[103] Wang, *supra note* 99, p. 211.

[104] *Ibid.*

[105] *Ibid.*

Code of Conduct for Responsible Fisheries as a supporting example.[106] In addition, the French delegate proposed that Taiwan be included in the IATTC and referred to the term "fishing entity" from the 1995 Fish Stock agreement.[107] Although the UNFSA used the term "fishing entities," Taiwan was still not sure of the status of "fishing entities" in RFMOs. Hence, Taiwan did not rule out the possibility of accepting the identity of a fishing entity but preferred the WTO model.

At the Third session of the Working Group in October 1999, the Chairman proposed a "Single Negotiation Text," which defined the term "parties" of the amended IATTC Convention as "the States, regional economic integration organisations and separate customs territories which have consented to be bound by this Convention and for which this Convention is in force."[108] Mexico suggested bracketing the definition of parties and discussion on it after solving other issues, which Colombia supported and the chairman agreement.[109] As a result, at the fourth session of the working group in May 2000, the chairman offered the "Revised

---

[106] *Ibid.* The Taiwanese delegate cited the example of Article 1.2 and Article 1.3 of the 1995 Code of Conduct for Responsible Fisheries. Article 1.2 states: "The Code is global in scope, and is directed toward members and nonmembers of FAO, fishing entities, sub regional, regional and global organisations, whether governmental or non-governmental, and all persons concerned with the conservation of fishery resources and management and development of fisheries, such as fishers, those engaged in processing and marketing of fish and fishery products and other users of the aquatic environment in relation to fisheries." Article 1.3 reads: "The Code provides principles and standards applicable to the conservation, management and development of all fisheries. It also covers the capture, processing and trade of fish and fishery products, fishing operations, aquaculture, fisheries research and the integration of fisheries into coastal area management."

[107] *Ibid.*

[108] *Ibid.*

[109] *Ibid.* The US delegate stated that the term "separate customs territory" was acceptable and presumed that the Mexican proposal was to reserve the issue for future discussion, rather than reject the term. The Mexican delegate agreed with the U.S. statement. Having diplomatic relations with Taiwan, Nicaragua noted that the term "separate customs territories" was used in several international instruments, such as the WTO Agreement, 1995 Grains Trade Convention and 1999 Food Aid Convention. The full text of the 1995 Grains Trade Convention is available at *http://ec.europa.eu/world/agreements/downloadFile.do?fullText=yes&treatyTransId=13582* (visited on 15/08/2012), and the full text of the 1999 Food Aid Convention at *http://www.foodaidconvention.org/Pdf/convention/iga1995.pdf* (visited on 15/08/2012).

Chairman's Text" defining "parties" as "the States, regional economic integration organizations [and separate customs territories] which have consented to be bound by this Convention and for which this Convention is in force."[110]

During the fourth meeting, Mexico suggested taking the MHLC as a reference as the MHLC and the IATTC both dealt with fisheries conservation and management in their respective regions.[111] It is supposed that Mexico would prefer that Taiwan participate under the identity of a fishing entity, instead of separate customs territories, because in the discussion at the MHLC, Taiwan agreed to join the WCPFC under the identity of a fishing entity. In response to the Mexican proposal, the US delegate noted that the two subregions to be governed by the WCPFC and the IATTC were in different geographical areas, involved different participants and had different histories; hence, similar considerations would not necessarily be involved in the designing of the two conventions.[112]

One week before the fifth session, it had become clear that Taiwan participated in the MHLC negotiation using the term "fishing entities," so many delegates in the working group preferred applying this model in the IATTC.[113] The situation did not favour Taiwan's participation under the WTO model. The WCPFC model was not the best result for Taiwan, so the Taiwanese delegate tried to win for the WTO model in the IATTC but was ready to accept the WCPFC model as a fall-back. At the fifth session of the working group in September 2000, the chairman proposed a "Main Pending Issues" paper, offering four definitions of the term "parties."[114] In the first definition, states, fishing entities and separate customs territories could equally become parties. This was the result that Taiwan had sought in the IATTC; therefore, it was Taiwan's first choice. The second option copied the WCPFC model, creating the distinct concepts of convention parties and members of the commission. In addition, this definition replaced the term "separate customs territories" with the phrases "[entity] [fishing entity] [entity/fishing entity]." Under this option, Taiwan could become a member of the commission only as a fishing entity, not a convention party, resulting in an unequal status with them. The third

---

[110] Wang, *supra note* 99, p. 212.

[111] *Ibid.*

[112] *Ibid.*

[113] *Ibid.*, p. 213. The Taiwanese delegate also felt that Taiwan's participation in the IATTC could not supersede its participatory status in the WCPFC. *See* Ho, *supra note* 21, p. 143.

[114] The content of these four options is discussed in Chapter 3 of this book.

option was derived from the ICCAT model, in which Taiwan, as a fishing entity, was restricted to being an observer. The last definition suggested adopting the rules of general international law and the decisions of the commission, resulting inflexibility and ambiguity. This option could be the most disadvantageous for Taiwan as it might be limited to the traditional concept of state, leading to the rejection of Taiwan's participation even as a fishing entity. For Taiwan, the second option was not fully satisfactory but still acceptable as it could at least possess the same status as in the WCPFC; however, the last two options were not acceptable or feasible solutions.[115]

At the Sixth session of the working group, the chairman provided the "Chairman's Consolidated Text"[116] which corresponded to the second option, the WCPFC model. In an informal talk with Taiwan, the chairman explained this model and that it was specially designed for Taiwan to avoid political obstacles. He suggested that continuing to struggle for other rights with political implications might not favour Taiwan's interests in the IATTC.[117] In the Chairman's Consolidated Text, the term "separate customs territories" does not appear in the definition of either parties or members of the commission. However, Article Blank regulates the participation of "[a]ny [separate customs territory] [entity] [fishing entity] [entity/fishing entity] whose vessels fish for highly migratory species in the Convention Area."[118]

---

[115] Wang, *supra note* 99, p. 213.

[116] Article I of the "Chairman's Consolidated Text" defines "parties" as "the States and regional economic integration organisations which have consented to be bound by this Convention and for which this Convention is in force, in accordance with the provisions of Articles XXVI, XXVII, XXVIII, and XXXIX of this Convention," "Members of the Commission" refers to "the Parties and any [separate customs territory] [entity] [fishing entity] [entity/fishing entity] which has expressed in accordance with the provisions of Article.. of this Convention its formal commitment to implement and comply with this Convention and any conservation and management measures adopted pursuant thereto." *See ibid.*, p. 214.

[117] *Ibid.*, p. 213.

[118] The full text can be seen in Wang, *supra note* 99, p. 214; and Hu, *supra note* 37, p. 170. Wang argues that Taiwan does not believe that it fits this definition because juxtaposing "entity" and "fishing entity" leads to the implication that "fishing entity" has the same standing as "entity," which in UNCLOS means self-governing associated states or territories which enjoy full internal self-government but have not attained full independence. Therefore, Taiwan avoids using the term "entity" to refer to itself because it believes that "entity" carries a connotation of being something other than a state. It prefers the term "fishing entity" and supports the distinction between these terms. However, in Serdy's view, "fishing entity" is a

In the opening speech of the seventh session, the chairman expressed support for Taiwan's participation, arguing that it would increase the effectiveness of the conservation and management fishery in the IATTC. He further noted that "the IATTC was an independent regional organization in that there was no linkage between the IATTC and the United Nations. As a consequence, the issue of incorporating Taiwan's participation in the work of the IATTC had to be discussed independently from political issues connected with the United Nations." [119] Most delegates agreed with the chairman's speech, but Mexico, which spoke for the PRC, indicated that it would not accept Taiwan's participation, whether as a party or as a member of the commission.[120] The Taiwanese delegate responded that "Taiwan had tried to avoid certain political considerations through promoting the designation of 'separate customs territory' or 'fishing entity' rather than statehood. For Taiwan, seeking status as a 'Separate Customs Territory' or 'Fishing Entity', rather than statehood, demonstrated that as one of the major distant-water fishing nations in the world Taiwan was committed to circumventing sensitive political barriers and to work with members of the IATTC on the long-term conservation and the sustainable utilization of the tuna stocks." [121] So far, the atmosphere was favourable to Taiwan's participation, so the issue in the future would be what identity and name Taiwan would use.[122]

In addition, it is worth noting that the PRC requested to join as an observer in the eighth session of the Working Group and received a positive response from the chairman. [123] At the eighth session, the definition of "members of the commission" in the "Chairman's Consolidated Text" had been changed, removing the bracketed options and leaving only "fishing entity" without brackets. During this session, the PRC reiterated its One China Policy, firstly asking to change the accepted practice of the table setting; it requested to remove all name boards and

---

subset of "entity." Both terms indicate an actor which is different from a state. In this case, the author holds that, if Taiwan prefers to emphasise that "entity" and "fishing entity" are both non-state actors but have different meaning in RFMOs, then putting "entity," "fishing entity" or other terms together implies that each concept has different meaning and explanations which would not limit Taiwan's status inside or outside the RFMOs. Also *see* Serdy, *supra note* 68, pp. 194-195.

[119] The original document is not accessible by the public, so the chairman's speech is recounted from the article by Wang, *supra note* 99, p. 215.

[120] *Ibid.*

[121] *Ibid.*

[122] *Ibid.*

[123] *Ibid.*

national flags from the meeting table.[124] The request was not met, whereupon the PRC delegate declared that:

> Taiwan does not have the qualifications to join [IATTC], neither should it be admitted to the Meetings as an observer in the name of Taiwan. … China suggests the settlement of the Taiwan issue in other regional fisheries management organizations be followed.[125]

In response to the PRC, Taiwan argued that:

> It is not a deviation to see all regional economic integration organizations, members of WTO, or fishing entities to become Contracting Parties to the IATTC. Rather, having an inclusive, comprehensive regime with a universal participation of all active actors in this regional fisheries management organization secures the effectiveness of the Convention and the organization themselves.[126]

Some states with diplomatic relations with Taiwan declared their support for its participation, and the United States stated that Taiwan's participation would not conflict with the US-defined "One China Policy."[127] Then, the PRC delegate proposed a MHLC Annex approach, which suggested adopting the WCPFC formula and including Taiwan's participation in the Annex of the Convention.[128] China's proposal was rejected by some delegates and the chairman, who did not want to copy the WCPFC Convention experience.[129] Furthermore, at the Eighth session, the chairman asked Taiwan if would accept membership status under the identity of fishing entity, and the Taiwanese delegate stated that:

> It will be acceptable provided if Taiwan can be assured of obtaining "full membership and equal participation."[130]

Before the ninth session, the chairman offered a Revised Consolidated Text, in which Article I defined "parties" and "members of the commission," confirming that the WCPFC model would be the final version. The definition of fishing entity given in Article I and the

---

[124] Hu, *supra note* 37, p. 167.
[125] Wang, *supra note* 99, p. 215.
[126] *Ibid.*; for more details of Taiwan's response, *see* Hu, *supra note* 37, pp. 167-168.
[127] Hu, *supra note* 37, p. 168.
[128] *Ibid.*, p. 170.
[129] *Ibid.*
[130] *Ibid.*, p. 169.

provisions concerning its participation were regulated in Article XXVIII. It was clear that Taiwan's participation would follow the WCPFC model, making it a member of the commission, not a party. However, the issues concerning the status of fishing entities within the IATTC were still not settled. The chairman insisted on the "transparency" principle in dealing with the difficult political issue between Taiwan and the PRC; therefore, the chairman hosted an informal, closed-door tripartite meeting between Taiwan and the PRC during the ninth and 10th sessions of the working group.[131] These unofficial and unprecedented meetings offered an opportunity for the two sides to negotiate their important interests concerning RFMOs face to face. During the tripartite meeting, Taiwan concentrated on the substance of fishing entities' status in the IATTC, while the PRC objected to Taiwan possessing at least two rights: involvement in the depositary procedure and being elected as the chair or a vice-chair of the commission.[132] Additionally, the PRC proposed a list of the names which it found acceptable for Taiwan's participation: Taiwan, Province of China; Taiwan, China; Chinese Taiwan; Chinese Taipei; and Taipei, China.[133] Taiwan finally accepted Chinese Taipei as its designation because that name had been used in other international organisations.

At the 70th IATTC Meeting in June 2003, the draft of the Antigua Convention completed by the working group was adopted. Taiwan, with the identity of a fishing entity and the designation of Chinese Taipei, signed an Instrument for the Participation of Fishing Entities in Washington on 14 November 2003 and become a member of IATTC once the Antigua Convention entered into force on 27 August 2010.[134] Although neither Taiwan nor the PRC were satisfied with the result, it was the best approach that could be accepted by both.

Notwithstanding Taiwan's membership in the IATTC, its conflict with the PRC has not ceased. Before the 81st Meeting of the IATTC, the first after the Antigua Convention entered into force, the PRC sent the Secretariat a letter which restated its One China Policy and requested some arrangement be made by the commission:

> 1. The use or appearance of the so-called "National Flag," "National Emblem" or "National Anthem" of Taiwan authority is forbidden.

---

[131] *Ibid.*, p. 171.

[132] Wang, *supra note* 99, p. 216.

[133] *Ibid.* According to Wang's observation, Taiwan would not accept any item of the list because it wanted to avoid creating the impression that it was being dictated to by China on the issue of designation.

[134] *See* Chapter 3, fn 49, of this book.

> 2. Chinese Taipei, as a fishing entity, should be arranged to be seated after the participating States. In the meeting documents, inter alia, the name list of participants, Chinese Taipei should also be listed after the participating States.
> 3. In all meeting activities, publications, documents, materials and other relevant items, the use of any terms that have sovereign implication such as "Republic of China," "R.O.C," "Taiwan," "Ministry of Foreign Affairs" and "Executive Yuan" etc. is forbidden.[135]

In response to the PRC's request, Taiwan also sent a letter that reiterated its concessions to the IATTC and opposed any political manoeuvring in the IATTC:

> [T]he adoption of the Antigua Convention itself is an achievement of collective wisdom by the original drafters and negotiators. Taiwan made tremendous concessions during the negotiations that led to the adoption of the Antigua Convention by, for instance, agreeing to use "Chinese Taipei" as its designation.
>
> IATTC is a [RFMO] … not a political forum to settle so-called "one China issue." We do not like to see such political maneuvering made by the Chinese Delegation occurring in a body like IATTC. Such political maneuvering will not contribute to the work of the IATTC but will only damage the cooperative spirit shown in the past negotiations and the goals of the IATTC in the years to come.
>
> As a member of the IATTC, … Chinese Taipei cannot accept any unequal treatment in sitting arrangement or presentation in any activities, publications, documents, materials of the IATTC, suggesting any differentiation from any other members of the IATTC.[136]

Ironically, while at MHLC6 in 2000, Taiwan had mentioned its presidential election and emphasised its democratic development in democracy, and China opposed politicising fishing issues.[137] However, in the IATTC arena, they reversed positions on politicisation. This dispute

---

[135] The full text of PRC's letter can be found in the Report of the IATTC 81st Meeting, Appendix 5a, pp. 96–97, available at *http://www.iattc.org/Meetings/Meetings2010/PDF/Sept/Minutes/IATTC-81-Sep-2010-Minutes.pdf* (visited on 18/12/2012).

[136] The full text of Taiwan's letter can be found in the Report of the IATTC 81st Meeting, Appendix 5b, pp. 98–100.

[137] *See* the relevant statements from Taiwan and the PRC in MHLC6, *supra note* 68 and 69.

again revealed the difficulty of dealing with the irreconcilable positions of China and Taiwan.

After consultation, the chair announced that the commission did not agree to China's requests regarding meeting arrangements.[138] In response, China stated that, under these circumstances, China could not agree to any formal resolutions offered for adoption.[139] Several delegations urged China to be more flexible, but China stated that it could not modify its position on formal resolutions, given the importance for China of the issue of meeting arrangements relative to Chinese Taipei. Eventually, China clarified that, although it would not agree to formal commission resolutions, it would not object to less formal recommendations that other members wished to pursue.[140] In fact, no evidence shows that China engages in further activity to affect the IATTC's functioning in practice.

4.3 International Scientific Committee for Tuna and Tuna-like Species in the North Pacific Ocean

In the first meeting of the ISC in May 1996, Taiwan was invited as an observer under Part B of the ISC Guidelines. These specified that members are "coastal states of the region and states with vessels fishing for tuna and tuna-like species in the region" and that observer participants are "relevant intergovernmental fishery organisations, relevant intergovernmental marine science organisations, and other entities with vessels fishing for tuna and tuna-like species in the region."[141] While discussing the issue of the "expansion of participants," Taiwan expressed its interest in becoming a member, instead of an observer participant. Taiwan stated that "we would like to see collective and cooperative efforts and respectable participation on a non-discriminatory basis in any regional or subregional fisheries organization or arrangement."[142]

Subsequently, Taiwan proposed an amendment to the guidelines during the second meeting of the ISC in January 1999.[143] It modified Part B of the

---

[138] *See* the Report of the IATTC 81st Meeting, *supra note* 135, p. 4.
[139] *Ibid.*
[140] *Ibid.*
[141] *See* the Report of the First Meeting of the ISC on official ISC website, *http://isc.ac.affrc.go.jp/pdf/ISC1pdf/isc1P_rep.pdf* (visited on 12/06/2012).
[142] The statement was recorded in Attachment 5 of the Report of the first meeting, *see ibid.*
[143] *See* the statement in Appendix 1 of the Report of the Second Meeting of the ISC, pp. 34–35, 39–40, *http://isc.ac.affrc.go.jp/pdf/ISC2pdf/isc2P_rep.pdf* (visited on 12/06/2012).

guidelines, deleting the item concerning other entities with vessels fishing for tuna and tuna-like species in the region in the part of observer participants and adding "economies of the region" and "economies with vessels fishing for tuna and tuna-like species in the region" in the part on members (see Table 7.2).[144] In other words, Taiwan tried to participate as a full member under the identity of "economy," following the APEC model. The assumption was that as APEC members, all the ISC members would be willing to accept the concept of economy. As, well under the APEC, each economy has equal status, which implied that Taiwan was equal to other ISC members even though it participates under the identity of economy rather than state.[145] The United States declared that it could accept this proposal provided that the ISC was a technical body and did not take legally binding actions.[146] The Japanese delegation stated that this issue should be resolved through mutual understanding among members.[147] In addition, the PRC expressed its opposition, arguing that the ISC was still in the developmental stage and not at the time to extend its membership.[148] The PRC proposed a slight modification in the part on observer participants, changing "other entities with vessels fishing for tuna and tuna-like species in the region" to "other entities, fishing entities with vessels fishing for tuna and tuna-like species in the region (see Table 7.2)." This change implied that the PRC would like Taiwan to join as an observer rather than a member.[149] Responding to the PRC proposal, Taiwan contended that its proposal based on the APEC formulation was a more appropriate solution.[150] The chair suggested that "the delegations should consult with higher authority in their governments with regard to this issue and that consultation between governments should be encouraged so that the issue can be solved before the next ISC meeting."[151] The amendment was not adopted.

Since this amendment was not adopted, Taiwan proposed another (see Table 7-2) during the third meeting in 2002. This amendment would allow fishing entities to be qualified directly as members rather than observer

---

[144] *Ibid.*
[145] Hu, *supra note* 37, p. 174.
[146] *See* the statement in Appendix 1 of the Report of the Second Meeting of the ISC, *supra note* 143, p. 35.
[147] *Ibid.*
[148] *Ibid.*
[149] *Ibid.*
[150] *Ibid.*
[151] *Ibid.*

participants.[152] This proposal indicated that Taiwan had given up on the APEC model and decided to participate under the identity of fishing entity. The chair accepted Taiwan's request to circulate its new version of the Guidelines to all participants before the adjournment of the meeting.[153] The proposal was adopted, making Taiwan able to become a member of the ISC.[154] During the meeting, the Taiwanese delegate expressed its government's willingness to join the ISC:

> After deliberation, we have come to a conclusion to amend the existing Guidelines to incorporate Taiwan as a full Member of the ISC. I trust that this amendment reflects the general recognition by all of the participants of the ISC of Taiwan's contribution in this body. I am encouraged by this latest development, and I would like to share with you that Taiwan will regard itself as a constructive force in this body in the days to come.[155]

Interestingly, the PRC did not attend this meeting.[156]

5. Conclusion

The promulgations of the territorial sea and EEZ by the Taiwan government did not encounter the opposition or protest from the PRC. In geographical strategy, the PRC clearly has adopted the strategy of maintaining the status quo, rather than triggering a war. Under the One China Policy, Taiwan as a *de facto* state is still able to participate in the international community. Although the PRC strongly prevents Taiwan from being recognised as a state, it seldom opposes bilateral treaties between Taiwan and other states. Even the fishery agreement concerning the overlapping EEZs between Taiwan and Japan did not face opposition from the PRC. However, in multilateral treaties, the PRC strongly opposes Taiwan becoming a contracting party even under a different identify than a state. A possible reason for China's opposition is that involvement in the process of negotiating, preparing, drafting and signing a multilateral treaty could help Taiwan to increase its visibility in international forums, which

---

[152] *See* the Report of the Third Meeting of the ISC on the official ISC website, pp. 14, 23–24, *http://isc.ac.affrc.go.jp/pdf/ISC3pdf/isc3P_rep.pdf* (visited on 12/06/2012).
[153] *Ibid.*
[154] *Ibid.*
[155] *Ibid.*, p. 30.
[156] *See* Appendix 3, List of Participants, in the Third Meeting Report, *supra note* 152, pp. 18–22.

is one of Taiwan's diplomatic strategies. One way for the PRC to stop Taiwan's strategy is to block Taiwan from concluding multilateral treaties. With respect to a bilateral treaty, it is reasonable for the PRC not to recognise but rather to ignore the treaty between Taiwan and the other state. However, many states are involved in a multilateral treaty, so the PRC cannot unilaterally deny the existence or legality of the treaty. The most important reason might be that in a multilateral treaty, the PRC and Taiwan might both become contracting parties, which implies that the PRC and Taiwan have the same status vis- à-vis other states. This could support the impression that Taiwan is a *de facto* state and work against the PRC's claim that Taiwan is a part of China and rightfully controlled or ruled by the PRC. Moreover, although the majority of states do not *de jure* recognise Taiwan as a state, the issue over whether Taiwan is a state will likely be raised in a multilateral treaty, and Taiwan tends to be regarded as a *de facto* state by the contracting parties though it does not sign or accede to the treaty under the identity of a state.

The diplomatic competition between Taiwan and the PRC affects Taiwan's participation in RFMOs. Taiwan never ceases to maintain that it is a state with the name of ROC. However, given the events of history, Taiwan is forced to participate in RFMOs as a fishing entity based on its functional aspect rather than the status of a state. Throughout its participation in RFMOs, Taiwan has shown flexibility by using different names and the identity of a fishing entity. Although unsatisfied, Taiwan successfully became a member of the WCPFC, IATTC and ISC. However, Taiwan failed to acquire the status of contracting parties, even members, in some RFMOs. Taiwan's participation in those RFMOs is discussed in the following chapter.

Table 7-2

Original Provision of Part B in the ISC Guidelines	**Amendment Proposed by Taiwan (in the 2nd Meeting, not adopted)**	**Amendment Proposed by the PRC**	**Second Amendment Proposed by Taiwan (adopted)**
B. Membership 1. Member a. Coastal states of the region b. States with vessels fishing for these species in the region 2. Observer participants a. Relevant intergovernmental fishery organisations b. Relevant intergovernmental marine science organisations c. Other entities with vessels fishing for those species in the region	B. Membership 1. Member a. Coastal states/ **Economies** of the region b. States/ **Economies** with vessels fishing for these species in the region 2. Observer participants: a. Relevant intergovernmental fishery organisations b. Relevant intergovernmental marine science organisations c. **(Delete)**	B. Membership 1. Member a. Coastal states of the region b. States with vessels fishing for these species in the region 2. Observer participants a. Relevant intergovernmental fishery organisations b. Relevant intergovernmental marine science organisations c. Other entities**, fishing entities** with vessels fishing for those species in the region	B. Membership 1. Member a. Coastal states/ **fishing entities** of the region b. States/ **fishing entities** with vessels fishing for these species in the region 2. Observer participants a. Relevant intergovernmental fishery organisations b. Relevant intergovernmental marine science organisations c. **(Delete)**

Source: Reports of the First, Second and Third Meetings of the ISC, *http://isc.ac.affrc.go.jp/reports/index.html* (visited on 12/06/2012).

CHAPTER EIGHT

QUASI-MEMBERSHIP, OBSERVER AND INVITED EXPERT

1. Introduction

Taiwan, as a fishing entity, participates as a member in RFMOs such as the WCPFC, IATTC, and ISC as discussed in chapter seven. However, Taiwan cannot become a member in some RFMOs, including the CCSBT and ICCAT, due to constraints in their regulations. For Taiwan's participation, CCSBT established an Extended Commission[1] which largely replaced the functions of the original commission. Although Taiwan is an observer in the CCSBT, it can be involved in the whole commission as a member of the Extended Commission; therefore, its status here is that of a quasi-member. In the ICCAT, Taiwan changed from an observer to a cooperating fishing entity, which helped normalise[2] its status. In addition, the IOTC stands as a special example because Taiwan participates not as a fishing entity but through various individuals selected as invited experts. The IOTC almost fully excludes Taiwan from participation as it falls inside the FAO framework and is the only tuna RFMO that cannot find a solution for Taiwan's participation. In recent years, the IOTC unsuccessfully tried to rid itself of FAO control so as to bring Taiwan into its conservation and management measures. Following the discussion of Taiwan's participation in RFMOs as a member, this chapter focuses on the process by which Taiwan gained special status other than a member in the CCSBT, ICCAT and IOTC.

---

[1] Regarding the Extended Commission of the CCSBT, *see* Chapter 3 of this book.

[2] Before using the identity of a cooperating fishing entity, Taiwan participated in the ICCAT as an observer with the identity of a State. However, China opposed this status. Thus, the identity of cooperating fishing entity provided a way for both Taiwan and China to accept Taiwan's participation in the ICCAT.

2. Taiwan's Participation in RFMOs as a Quasi-Member, Observer and through Invited Experts

2.1 Commission for the Conservation of Southern Bluefin Tuna

The CCSBT was established in 1994 with the purpose of ensuring the conservation and optimum utilisation of SBT.[3] The catch of Taiwan's vessels fishing for SBT increased from 80 tonnes in 1969 to 1920 tonnes in 1994, which ranked as the third largest catches at that time.[4] It is important for the CCSBT to invite Taiwan to participate in its conservation and management procedure. However, it is unlikely that Taiwan may directly apply to accede to the 1993 CCSBT Convention. Under Article 18 of the 1993 convention, the convention is only open to states to accede, and the existing parties recognise the PRC and its One China Policy in which it claims sovereignty over Taiwan. Therefore, it seems that this problem could be solved if Taiwan participated under another identity, such as a fishing entity, rather than a state. However, this would require amending of Article 18 of the 1993 convention to allow accession by fishing entities, and the process of amending a multilateral treaty mighty cost a long time because it needs the approval of each party's parliament.[5] Hence, amending the 1993 convention faces a certain degree of difficulty.

Pursuant to Article 14(1) which states that "any State or entity not party to this Convention, whose nationals, residents or fishing vessels harvest southern bluefin tuna" may be invited to send observers to meetings of the commission and Scientific Committee, Taiwan has participated as an observer since the first CCSBT meeting in 1994. The commission then noted Taiwan's desire to have greater participation than as merely an observer.[6] However, since no party recognised Taiwan, the

---

[3] Regarding the objective of the convention, *see* Article 3 of the CCSBT Convention. SBT refers to southern bluefin tuna, as defined in Chapter 3 of this book.

[4] In 1994, the largest catch was by Japan, with 6063 tonnes; and the second was by Australia, with 4700 tonnes. About estimates of the global SBT catch, *see* "Global Southern Bluefin Tuna Catch By Flag" on the official CCSBT website, *http://www.ccsbt.org/site/sbt_data.php* (visited on 15/08/2012).

[5] *See* Andrew Serdy, "Bringing Taiwan into the International Fisheries Fold: The Legal Personality of a Fishing Entity," in James Crawford and Vaughan Lowe, eds., *The British Year Book of International Law 2004* (Oxford: Clarendon Press, 2005, Vol. 75), p. 191.

[6] *Ibid.*, p. 188.

CCBST considered establishing the co-operation relation with Taiwan. At the First Special Meeting in 1996, it stated:

> It was agreed that active communication with non-parties was important and the Commission agreed that the Chair would write to .... Taiwan offering to engage in discussions regarding ... co-operation with the Convention. Following this, consideration would be given to representatives from Commission members visiting their fisheries or other authorities to encourage participation in the Commission, or in Taiwan's case, given that none of the parties currently recognises Taiwan, close co-operation with Commission initiatives.[7]

Taiwan might have been aware of the commission's intention; hence, in the Third Meeting of the CCSBT in September 1996, it agreed to restrict future catches to no more than the 1995 level of 1447 tonnes, starting in 1996.[8] In addition, since the UNFSA had been adopted in 1995, Taiwan's observer noted Taiwan's rights to join regional management regimes under it.[9] Taiwan did not explicitly indicate which UNFSA articles granted this right. After the adoption of the 1995 agreement, the Taiwanese government considered whether it should accept the position of fishing entities and eventually decided to adopt it as the concept was still unclear and the acceptance of it might lead to the implication of Taiwan was giving up of statehood.[10] However, the first time that Taiwan used the identity of a fishing entity was in the MHLC2 in June 1997.[11] Hence, it

[7] *See* the Report of the First Special Meeting of the CCSBT, available at *http://www.ccsbt.org/userfiles/file/docs_english/meetings/meeting_reports/ccsbt_02/report_of_special_meeting1_part2.pdf* (visited on 15/08/2012). Also *see* the Report of the Fourth Annual Meeting of the CCSBT (the Second Session), available at *http://www.ccsbt.org/userfiles/file/docs_english/meetings/meeting_reports/ccsbt_04/report_of_ccsbt4_part2.pdf* (visited on 17/08/2012), p. 1.

[8] *See* the Report of the Third Annual Meeting of the CCSBT (First Session), Attachment K, available at *http://www.ccsbt.org/userfiles/file/docs_english/meetings/meeting_reports/ccsbt_03/report_of_ccsbt3_part1.pdf* (visited on 15/08/2012).

[9] *Ibid.*

[10] *See* Peter S.C. Ho, "The Impact of the Fish Stocks Agreement on Taiwan's Participation in International Fisheries Fora," *Ocean Development and International Law*, Vol. 37(2006), p. 144.

[11] It can be found in Taiwan's opening statement at the MHLC2. *See* Chapter 7 of this book.

can be supposed that Taiwan was referring to Articles 1(3), 8(3) and 17(3) of the UNFSA.[12]

Due to Taiwan's special political position in the international community, it still had difficulty becoming a member of the CCSBT despite self-restraint on catches. At the Fourth Annual Meeting in 1998, Japan's view shows that the CCSBT was considering formalisation of its cooperation relation with Taiwan.

> Japan stated that due to the sensitive nature of its relationship with Taiwan, it needed further time to decide on the appropriate form of an instrument. The position of the Japanese Government was to consider the above approach on its merits, incorporating political and legal concerns. It expressed its support for the Commission seeking cooperation with Taiwan, and undertook to provide general comments on the Commission's proposed approach, and the most appropriate type of instrument to use. … Japan suggested that the first step would be to seek a strong commitment from Taiwan that it would cooperate with the Commission.[13]

Taiwan acknowledged the possible direction of the CCSBT concerning its participation and argued that, since it agreed to limit its catches, it should have the same rights as other members. Taiwan further expressed its desire to become a full member of the CCSBT.[14] However, in the

---

[12] *See* Serdy, *supra note* 5, p. 189. Regarding the text of Articles 1(3) and 17(3) of the UNFSA, *see* Chapter 1 of this book. Article 8(3) of the UNFSA reads: "Where a subregional or regional fisheries management organization or arrangement has the competence to establish conservation and management measures for particular straddling fish stocks or highly migratory fish stocks, States fishing for the stocks on the high seas and relevant coastal States shall give effect to their duty to cooperate by becoming members of such organization or participants in such arrangement, or by agreeing to apply the conservation and management measures established by such organization or arrangement. States having a real interest in the fisheries concerned may become members of such organization or participants in such arrangement. The terms of participation in such organization or arrangement shall not preclude such States from membership or participation; nor shall they be applied in a manner which discriminates against any State or group of States having a real interest in the fisheries concerned."

[13] *See* the Report of the Fourth Annual Meeting of the CCSBT (Second Session), *supra note* 7, pp. 1–2.

[14] In the opening statement at the Fifth Annual Meeting of the CCSBT, Taiwan's observer stated: "[O]bligation and right should be closely linked together. While performing obligation, we should have equal right as the same with other fishing nations. I would like to reiterate that Taiwan is very willing to cooperate with other countries for the conservation and management of marine living resources. Thus,

subsequent discussion with Taiwanese delegation, the CCSBT thought that the political reality should be taken into consideration and clearly stated that Taiwan's participation as a full member was not allowed by the 1993 convention.[15]

> The Commission had noted their interest in becoming a full member of the Commission. However, the Convention did not allow for this. They would also be aware of the political reality of this situation; the parties were not in a position to have Taiwan accede to the Commission with member status; [h]owever, the Commission wished to conclude an arrangement that provided for their cooperation with and participation in the work of the Commission. This would include the privileges and responsibilities that would accrue to a co-operating party.[16]

Taiwan's delegation responded that it was aware of the difficulties associated with accession to the convention but they hoped to learn the differences between the status of a cooperating party and a member and to be treated in an equitable manner with other members in the framework of the commission.[17] The CCSBT claimed that it acknowledged Taiwan's desire for equal participation and was looking to provide a way for it to have similar privileges and responsibilities as members.[18]

In October 1999, a CCSBT delegation visited Taiwan to discuss participation and fishing quotas.[19] The delegation suggested a quota of 1250 tonnes and status as a cooperating non-member subject to a contribution to the CCSBT's annual budget.[20] Taiwan refused this proposal because the offered fishing quota was far below its self-imposed limit of

---

we seek the accession to become a full member of CCSBT." *See* the Report of the Fifth Annual Meeting of the CCSBT (First Session), Attachment 9, available at *http://www.ccsbt.org/userfiles/file/docs_english/meetings/meeting_reports/ccsbt_05/report_of_ccsbt5_part1.pdf* (visited on 15/08/2012).

[15] However, per Serdy's view, Article 18 of the 1993 convention only concerns the accession of a state, so the real obstacle to Taiwan's becoming a full member is not the convention itself but the political conflict between Taiwan and the PRC, which creates the recognition issue among the parties, PRC and Taiwan. *See* Serdy, *supra note* 5, p. 189.

[16] *See* the Report of the Fifth Annual Meeting of the CCSBT (Second Session), Attachment G, available at *http://www.ccsbt.org/userfiles/file/docs_english/meetings/meeting_reports/ccsbt_05/report_of_ccsbt5_part2.pdf* (visited on 15/08/2012).

[17] *Ibid.*

[18] *Ibid.*

[19]. Ho, *supra note* 10, p. 141.

[20] *Ibid.*

1450 tonnes. Taiwan considered that, if its status was as a non-member, then it might agree to make contributions or donations voluntarily but could not accept the obligation to contribute to the organisation's budget being imposed on it.[21] Although the negotiation was not successful, it is noted that Taiwan could accept participating as a cooperating non-member. The problem that still needed to be resolved was the quota and other details.

However, at the Sixth Annual Meeting in November 1999, Taiwan made a surprise move, delivering a letter applying for becoming a party to the 1993 convention and stating that, "if an arrangement is needed to make our Party status possible, we would only accept an arrangement that is fair, workable, and duly respecting Taiwan's status."[22] While analysing Taiwan's application, Andrew Serdy writes:

> In doing so it should be noted that Taiwan was not purporting to accede to the 1993 Convention; this would have required lodging an instrument of accession with Australia as depositary. It can, however, be taken as an implied request to the Parties to amend the Convention so as to allow entities other than States to accede. Taiwan may have had in mind the first draft of the new convention being negotiated in the [IATTC].[23]

Indeed, during negotiations in 1999, Taiwan was aware of and seemed willing to accept the CCSBT's position that Taiwan should be a cooperating non-member, but still applied for accession. Taiwan knew that its application would not be accepted, so its purpose might have been to put on record its application for the status of a state. Actually, as Taiwan hoped to accede to the convention with the identity of a state, its moves conform with Article 18 of the convention. Thus, the convention itself does not prohibit Taiwan from being a full member of commission. The real obstacle was the uncertain status of Taiwan due its conflict with China over recognition.[24]

In October 2000, the CCSBT sent another delegation to Taiwan, which proposed a draft resolution concerning the establishment of an Extended Commission, of which Taiwan could become a member.[25] The Extended

---

[21] *Ibid.*

[22] *See* the Report of the Sixth Annual Meeting of the CCSBT (First Session), Attachment H, available at *http://www.ccsbt.org/userfiles/file/docs_english/meetings/meeting_reports/ccsbt_06/report_of_ccsbt6_Part1.pdf* (visited on 17/08/2012).

[23] Serdy, *supra note* 5, p. 190.

[24] *Ibid.*, p. 189.

[25] Ho, *supra note* 10, p. 141.

Commission would meet annually before the CCSBT annual meeting and perform nearly the same function as the commission, such as deciding upon a total allowable catch and its allocation among the members. Then the CCBST proper would adopt and implement the decisions of the Extended Commission while reserving the right to reject those decisions. Taiwan believed that this resolution lacked a concrete legal foundation for the establishment of the Extended Commission and so could be easily revoked or replaced by another resolution.[26] In addition, Taiwan was concerned that accepting the proposal would undermine its attempts to participate in other RFMOs as a contracting party or member.[27] The CCSBT delegation responded that, considering the existing political environment, the proposal was the best compromise possible. The CCBST assured that the commission would serve as a rubber stamp for of the Extended Commission's decisions unless they were highly controversial. The CCBST also argued that after the resolution was adopted, it would be accepted by any new members of the CCSBT.[28]

At the Seventh Annual Meeting, Taiwan referred to the visiting CCSBT delegation and clearly expressed its desire to achieve equal status with other members:

> In various occasions, we have repeatedly expressed that obligation of a fishing nation should be commensurate with its right. We have also expressed that we have the strong willingness to participate as a member on an equal basis as other members in CCSBT. CCSBT organized a delegation to visit Taipei last October, and thoroughly exchanged views with us on a new proposal presented by CCSBT to accommodate Taiwan in the regime. … My government is prudently evaluating the CCSBT's proposal because of its complicated nature and involvement in our domestic legal and political circumstances. What Taiwan intends to secure

---

[26] *Ibid.*

[27] *Ibid.*

[28] *Ibid.* Pursuant to Article 8(7) of the CCSBT Convention, "all measures decided upon under paragraph 3 above shall be binding on the Parties." The "Parties" to which Article 8(7) refers include any new entrants which accede to the Convention; thus, a new entrant to the CCSBT is bound not only by the convention itself but also by past decisions. This is what Serdy called "Acquis Commissionaire." For further details, *see* Andrew Serdy, "Postmodern International Fisheries Law, or We Are All Coastal States Now," *The International and Comparative Law Quarterly*, Vol. 60(April 2011), pp. 400–404.

> is an equal status and full participation as other parties to the Convention in the CCSBT regime.[29]

Then, during the same meeting, the Resolution to Establish an Extended Commission and an Extended Scientific Committee was adopted.[30] Taiwan expressed its willingness to become a member of the Extended Commission.[31]

At the Eighth Meeting of the Commission in October 2001, Taiwan stated that it would apply to be a member of the Extended Commission and the Extended Scientific Committee for CCSBT under the name of either Chinese Taipei or the Fishing Entity of Taiwan as soon as possible.[32] The executive secretary responded that:

> Taiwan should lodge a formal application for Membership of the Extended Commission by an Exchange of Letters as provided for in Paragraph 6[33]of the Resolution to Establish an Extended Commission and Extended Scientific Committee … by 31 December 2001. The Commission will correspond with Taiwan shortly to initiate this process.[34]

Then, at the Ninth Meeting in October 2002, the chair announced that Taiwan's membership in the Extended Commission as a fishing entity

---

[29] *See* the Opening Statement by Taiwan in the Report of the Seventh Annual Meeting of the CCSBT, Attachment D-5, available at *http://www.ccsbt.org/userfiles/file/docs_english/meetings/meeting_reports/ccsbt_07/report_of_ccsbt7.pdf* (visited on 19/08/2012).

[30] *Ibid.*, Attachment I.

[31] *See* the Report of the Seventh Annual Meeting of the CCSBT.

[32] *See* Statement made by Taiwan in the Plenary in the Report of the Eighth Annual Meeting of the Commission, Attachment F, available at *http://www.ccsbt.org/userfiles/file/docs_english/meetings/meeting_reports/ccsbt_08/report_of_ccsbt8.pdf* (visited on 19/08/2012).

[33] Paragraph 6 of the Resolution to Establish an Extended Commission and Extended Scientific Committee reads: "Any entity or fishing entity, vessels flagged to which have caught SBT at any time in the previous three calendar years, may express its willingness to the Executive Secretary of the Commission to become a member of the Extended Commission. The Executive Secretary of the Commission, on behalf of the Commission, will conduct an Exchange of Letters with the representative of such entity or fishing entity to this effect. In so doing, the applicant shall give the Commission its firm commitment to respect the terms of the Convention and comply with such decisions of the Extended Commission as become decisions of the Commission pursuant to paragraph 4."

[34] The full text can be found in the Report of the Eighth Annual Meeting of the Commission, *supra note* 32.

would became effective on 30 August 2002.[35] Therefore, Taiwan has two identities in the CCSBT: a member of the Extended Commission and an observer in the CCBST proper. To avoid the possibility of being forced to accept a decision adopted without Taiwan's agreement, Taiwan wrote a note to the CCSBT Secretariat.

> The Fishing Entity of Taiwan understands that in accordance with the Rules of Procedure of the Commission and Extended Commission, all decisions must be taken by a unanimous vote of Members present. Furthermore, any decision made by the Commission or any of its subsidiary bodies on an issue on which unanimity in the Extended Commission has not been achieved, or any issue that has not been deliberated in the Extended Commission, shall not be binding on Members of the Extended Commission which are not Members of the Commission.[36]

Despite knowing that there was almost no possibility of becoming a full member of the CCBST, Taiwan was still dissatisfied with its status in the CCSBT and hoped to become a full member.[37]

It is worth noting that Taiwan's agreement to accept this resolution came after the adoption of the WCPFC Convention, to which Taiwan was not allowed to accede as a contracting party but, rather, as a full member. This formula set a better precedent than the CCSBT Extended Commission for Taiwan's participation in other RFMOs.[38] Hence, the possibility of the consequence that had concerned Taiwan—creating a precedent that would be an obstacle to joining other RFMOs—was reduced. However, it was obvious that, for Taiwan, the outcome of its participation in the CCSBT is not sufficient. In the WCPFC, Taiwan at least became a member even in the face of the PRC's challenge.

---

[35] *See* the Report of the Ninth Annual Meeting of the Commission, p. 1, available at *http://www.ccsbt.org/userfiles/file/docs_english/meetings/meeting_reports/ccsbt_09/report_of_ccsbt9.pdf* (visited on 19/08/2012).

[36] Ho, *supra note* 10, p. 142.

[37] *See* Taiwan's Opening Statement at CCSBT9 and Statement by the Fishing Entity of Taiwan to the Admission in the Report of the Ninth Annual Meeting of the Commission, *supra note* 35, Attachment 5-2 and Attachment 1. Also *see* Serdy, *supra note* 5, p. 199.

[38] Ho, *supra note* 10, p. 142.

2.2 International Commission for the Conservation of Atlantic Tunas

Taiwan's involvement in the ICCAT was first recorded at the First Regular Meeting of the ICCAT which Taiwan did not attend but contributed an informal report entitled Taiwan's Tuna Fisheries and Tuna Fisheries Research, 1970 to the Standing Committee on Research and Statistics.[39] At that time, Taiwan was still recognised as a state and was recorded under its national name, the ROC. The next year, the UN adopted Resolution 2758, which gave Taiwan's seat to the PRC.[40] Hence, Taiwan could participate in the ICCAT as a contracting party before 1971 but did not do so. Presumably, Taiwan did not foresee that its UN seat would be taken away soon. In addition, fisheries conservation and management was not yet a major concern of Taiwan. In 1972, the Taiwanese government began to attend the meetings of the ICCAT as an observer.[41] However, Taiwan's attendance was voluntary, so the obligations of continuing to attend the ICCAT annual meetings were not imposed on it.[42]

At the Thirteenth Regular Meeting in 1993, the ICCAT adopted Guidelines and Criteria for Granting Observer Status at ICCAT Meetings, which stipulated that non-contracting parties identified as harvesting tunas or tuna-like species in the ICCAT convention area should be invited as observers.[43] After the Guidelines, the PRC began to attend as an observer starting in the 1994 Ninth Special Meeting of the ICCAT.[44] During that meeting, the Resolution on Coordination with Non-Contracting Parties was adopted, urging non-contracting parties fishing in the convention area

---

[39] *See* the Report of the First Regular Meeting of the ICCAT, p. 49, 121, available at *http://www.iccat.int/Documents/BienRep/REP_EN_70-71_II.pdf* (visited on 22/08/2012).

[40] Regarding Resolution 2758 and the reason why the PRC replaced Taiwan's seat in the UN, *see* Chapter 7 of this book.

[41] *See* the Report of the Second Regular Meeting of the ICCAT, p. 37, available at *http://www.iccat.int/Documents/BienRep/REP_EN_72-73_I.pdf* (visited on 22/08/2012).

[42] On Taiwan's attendance at ICCAT meetings from 1972 to 1993, *see* Serdy, *supra note* 5, pp. 200–201.

[43] *See* the Report of the Thirteenth Regular Meeting of the ICCAT, Annex 19, pp. 137–139,available at *http://www.iccat.int/Documents/BienRep/REP_EN_92-93_II.pdf* (visited on 23/08/2012).

[44] *See* the Report of the Ninth Special Meeting of the ICCAT, Participant List, p. 69, available at *http://www.iccat.int/Documents/BienRep/REP_EN_94-95_I_1.pdf* (visited on 23/08/2012).

to become cooperating parties.[45] A cooperating party could also attend the ICCAT meetings as an observer. However, the main difference in observer and cooperating party status is the voluntary agreement to conform to the ICCAT decisions.[46] The US delegate encouraged Taiwan to become a cooperating party, and Taiwan expressed willingness to do so provided that its interests were dealt with on an equal basis.[47]

From the start, Taiwan regarded itself as a state attending the ICCAT Meetings as an observer. Even after the UN Resolution 2758 in 1971, Taiwan's statehood was still not mentioned by other contracting parties; neither did it become an issue in the ICCAT. So far, the PRC had not participated in the ICCAT, either as a contracting party or an observer. Taiwan's participation as an observer became an issue only after the PRC acceded to the ICCAT Convention in 1996, leading to the emergence of the term "fishing entities" in the ICCAT. During the Tenth Special Meeting of the ICCAT in 1996, a series of intense disagreements between the PRC and Taiwan occurred. The PRC protested the use of the name "Taiwan" by the ICCAT and objected that Taiwan was categorised as "non-member country" in some ICCAT documents, which did not accord with the PRC's One China Policy.[48]

> According to paragraph 3 of Article XI of the Convention[49] and Rule 5 of the Rules of Procedure[50], only international organizations and a government which is a member of the United Nations or any specialized agency of the United Nations and which is not a member of the Commission may be invited to send observers to the meeting of the

---

[45] *Ibid.*, Annex 10, p. 97.

[46] *See* paragraph 1 of the Resolution on Coordination with Non-Contracting Parties.

[47] *See* the Report of the Fourteenth Regular Meeting of the ICCAT, pp. 195–196, available at *http://www.iccat.int/Documents/BienRep/REP_EN_94-95_II_1.pdf* (visited on 23/08/2012).

[48] *See* the Report of the Tenth Special Meeting of the ICCAT, pp. 42-43, 52, 143, available at *http://www.iccat.int/Documents/BienRep/REP_EN_96-97_I_1.pdf* (visited on 24/08/2012).

[49] Article XI(3) of the ICCAT Convention reads: "The commission may invite any appropriate international organization and any Government which is a member of the United Nations or of any Specialized Agency of the United Nations and which is not a member of the Commission, to send observers to meetings of the Commission and its subsidiary bodies."

[50] Rule 5 of the Rules of Procedure has nearly the same stipulation as Article XI(3) of the ICCAT Convention but further states: "Observers may, with the authorization of the Chairman, address the meeting to which they are invited and otherwise participate in its work, but without the right to vote."

> Commission. Therefore, the invitation to and admission of Taiwan of China as an observer do not conform to the Convention and Rules of Procedure, and are also in violation of the Resolution 2758 (XXVI) of the United Nations General Assembly, and should be rectified immediately.[51]

In response to the PRC's statement, Taiwan explained its position:

> Taiwan expressed [its] hope that Taiwan and the [PRC] would be able to work together in friendly cooperation. He noted that this was the first time that a political issue of this nature had been raised at ICCAT. While the official name of his country is the name the "Republic of China," it has always accepted, as a compromise with ICCAT, the name of "Taiwan." The [ROC] has been in existence since 1912, thirty-eight years before the [PRC], and that this latter has unilaterally claimed the territory of Taiwan.[52]

The executive secretary explained that the secretary follows the established practices of accepting names that the non-contracting parties use, without any implication of recognition.[53] The executive secretary indicated that the ICCAT has no competence to solve international political problems.[54] In addition, the chairman of the ICCAT meeting stated that the ICCAT "felt the need to cooperate with Taiwan, as it has existed as a 'fishing reality' for many years."[55]

At the Fifteenth Regular Meeting of the ICCAT, Taiwan and the PRC made mutual concessions. Firstly, Taiwan as an observer changed its name to Chinese Taipei.[56] The ICCAT adopted the Resolution by ICCAT on Becoming a Cooperating Party, Entity or Fishing Entity,[57] which expands the category of non-contracting parties from "Cooperating Parties" in the 1994 resolution to "Cooperating Party, Entity or Fishing Entity." This was the first time that the ICCAT adopted the term "fishing entity," which was also the only position that both Taiwan and the PRC could accept. As the PRC stated:

---

[51] *See* the Statement by the People's Republic of China, in the Report of the Tenth Special Meeting of the ICCAT, *supra note* 48, at Annex 6-3, p. 101.
[52] *See* the Report of the Tenth Special Meeting of the ICCAT, *supra note* 48, p. 43.
[53] *Ibid.*, p. 143
[54] *Ibid.*
[55] *Ibid.*, p. 43
[56] There is no record explaining the process of how and why Taiwan decided to agree to change its name.
[57] On the source and other information about the resolution, *see* Chapter 3 of this book.

> We cannot accept Chinese Taipei being admitted to ICCAT meetings as an observer under the name of "Taiwan," which means splitting China and creates "one China, one Taiwan" within ICCAT.
>
> However, in order to help achieve the objectives of conservation, management and sustainable utilization of tuna resources in the Atlantic Ocean, and to take into consideration the fact that fishermen of Taiwan harvest tunas in the Convention area, the Chinese Delegation, in the spirit of cooperation … may accept Chinese Taipei attending ICCAT meetings as a fishing entity in the capacity of an observer under the designation of "Chinese Taipei."[58]

At first the Taiwanese government had considered not accepting being a fishing entity because it thought that it had conceded in agreeing to the designation of Chinese Taipei and did not want its statehood to be denied further.[59] Despite these considerations, Taiwan did agree to the position of a cooperating fishing entity.[60] One possible reason is that Taiwan had used the identity of fishing entity in MHLC2, which could be a precedent and overcome Taiwan's concerning about downgrading its statehood. The identity of cooperating fishing entity helped Taiwan move past the controversial issue of whether it should have observer status and normalise its status under the ICCAT Convention.

Despite becoming a cooperating fishing entity, Taiwan felt unsatisfied with this position compared to its status in other RFMOs. In the Twelfth Special Meeting of the ICCAT in 2000, Taiwan stated:

> Through years of cooperation with ICCAT, a resolution was finally adopted to enable us to attain Cooperating Status. For a party who is willing to become a member, but unable to do so, we see that the annual application for Cooperating Status is an underlying unfair treatment. Mr. Chairman, let me draw your attention to the recent development of a new [RFMO], which provides a mechanism for us to fully participate. … [I]t is our strong view that the arrangement of ICCAT should not be too far from that of the

---

[58] *See* the Report of the Fifteenth Regular Meeting of the ICCAT, Annex 6-2, p. 84, available at *http://www.iccat.int/Documents/BienRep/REP_EN_96-97_II_1.pdf* (visited on 24/08/2012).

[59] Ho, *supra note* 10, p. 138.

[60] *See* the Report of the Sixteenth Regular Meeting of the ICCAT p. 27, available at *http://www.iccat.int/Documents/BienRep/REP_EN_98-99_II_1.pdf* (visited on 24/08/2012).

> new fisheries convention. At least for the time being, we should not be required to apply annually for "Cooperating Status."[61]

Therefore, Taiwan's aim is to acquire the equal status as other full members. Surprisingly, there is no record concerning any opposition by the PRC to Taiwan's statement.

According to the 1997 resolution, the status of a cooperating party could not be continued automatically; rather, Taiwan had to apply for the status of cooperating fishing entity every year. To decrease this inconvenience, Taiwan in 2000 proposed amending the 1997 resolution to allow the status of the cooperating party, entity or fishing entity to be renewed automatically unless revoked by the commission due to non-compliance with ICCAT conservation and management measures or the commission received a written request for withdrawal of the status.[62] In addition, Taiwan suggested that a state eligible to become a contracting party could acquire the status of cooperating party, entity or fishing entity for no more than three years.[63] There is no record of why Taiwan proposed this provision. Taiwan might regard it as another opportunity to apply for the status of a member or a contracting party, but if its application was rejected, Taiwan would also fail to keep its normalised status of cooperating fishing entity and return to the uncertain status of an observer.

Taiwan's proposal was not adopted, but the next year, another draft entitled Resolution by ICCAT on Becoming a Cooperating Party, Entity or Fishing Entity was adopted to replace the 1997 resolution.[64] Under the 2001 resolution, Taiwan could automatically renew its status of cooperating fishing entity annually, but its suggestion that the status of cooperating party would be guaranteed for no more than three years was not included in the 2001 resolution. In 2003, the 2001 resolution was replaced again by a Recommendation on Criteria for Attaining the Status of Cooperating non-Contracting Party, Entity or Fishing Entity in ICCAT. The 2003 resolution kept most regulations in the 2001 resolution but listed

---

[61] *See* the Report of the Twelfth Special Meeting of the ICCAT, p. 72, available at *http://www.iccat.int/Documents/BienRep/REP_EN_00-01_I_1.pdf* (visited on 27/08/2012).

[62] The full text of Taiwan's proposal can be found in the Report of the Twelfth Special Meeting of the ICCAT, Appendix 2–Annex 10, pp. 251–252.

[63] *Ibid.*

[64] The full text of the 2001 resolution can be found in the Report of the Seventeenth Regular Meeting of the ICCAT, Annex 9-23, p. 261, available at *http://www.iccat.int/Documents/BienRep/REP_EN_00-01_II_1.pdf* (visited on 27/08/2012).

details about the documents that the cooperating party, entity or fishing entity must provide.[65] Although the ICCAT has never updated its Convention or the Rules of the Procedure, the resolutions and Recommendation concerning the cooperating party, entity or fishing entity, *inter alia* the 2003 recommendation, formalised the criteria to apply to be a cooperating fishing entity, providing Taiwan gained the legitimacy to participate in the ICCAT's operations.

2.3 Indian Ocean Tuna Commission—An Effective Tool for Sustainable Fisheries Development or Another Battlefield for Political Conflicts?

Since the late 1980s, Taiwan has been the major longline fleet fisher for bigeye tuna in the Indian Ocean, taking as much as 40% of the total longline catch.[66] Although its catches have decreased in recent years, Taiwan is still a major catcher of tuna in the Indian Ocean, so its participation is important for IOTC to effectively carry out its conservation and management measures. Under the IOTC Agreement[67], only UN members and specialised agencies, FAO members and associate members and members of the International Atomic Energy Agency can become a member of IOTC. In addition, the IOTC's Rules of Procedure stipulate that the non-state actors eligible to be observers are intergovernmental and non-intergovernmental organisations; thus, fishing entities can be observers only if they are also intergovernmental or non-intergovernmental organisations.[68] The IOTC falls under the UN/FAO framework which adopts the PRC's One China Policy viewing Taiwan a province of the PRC, so it is impossible for Taiwan to participate in IOTC as a state or under other identity equivalent to a state. As well, any approach for Taiwan to participate as a fishing entity is lacking.

---

[65] The full text of the 2003 resolution can be found in the Report of the Eighteenth Regular Meeting of the ICCAT, p. 167, available at *http://www.iccat.int/Documents/BienRep/REP_EN_02-03_II_1.pdf* (visited on 27/08/2012).

[66] *See* the Report of 15th Session of the IOTC Scientific Committee, p. 84, available at *http://www.iotc.org/files/proceedings/2012/sc/IOTC-2012-SC15-R[E].pdf* (visited on 29/05/2013).

[67] About the IOTC Agreement, *see* Chapter 3 of this book.

[68] The full text of the Rules of Procedure can be found on the official IOTC website, *http://www.iotc.org/English/info/rules_proced.php* (visited on 03/09/2012).

On 9 November 1999, Taiwan applied to the IOTC secretariat to become a full member and requested to be allowed to attend the Fourth Meeting of the IOTC as an observer government. The PRC strongly opposed this request, stating that it could only accept Taiwan's attendance at IOTC meetings as an observer from a non-intergovernmental organisation.[69] At the Fourth Meeting in 1999, the FAO Legal Adviser reported that

> the [PRC] accepted that a non-intergovernmental organization representing the fishing interests of Taiwan Province of China be invited to participate in IOTC meetings.[70]

Unfortunately, this issue was not discussed further. At the Sixth Meeting in 2001, Taiwan began to attend the IOTC meeting under the identity of invited expert.[71] Theoretically, invited experts are individuals and thus have no link to an official delegation. Therefore, the IOTC still needs a proper route to allow Taiwan to participate. On 2 September 2002, in accordance with Resolution 99/04 on the Status of Cooperating Non-Contracting Parties[72] adopted at the Fourth Meeting of the IOTC in 1999, Taiwan applied to become a co-operating non-contracting party. It was opposed by the PRC mainly due to the nomenclature that Taiwan hoped to use: Taiwan, not Taiwan Province of China, as the PRC insisted.[73]

In 2003, Taiwan's issue was discussed again but reached no result. With respect to Taiwan's participation, the PRC stated that:

> [I]t has shown great flexibility to adjust the concerns of all members regarding this issue. China is a responsible fishing State and will not leave the fishing fleets of Taiwan Province of China to operate in the Indian Ocean outside the IOTC. China is always willing to join other delegations

---

[69] *See* Huang-Chih Chiang, "On Taiwan's Participation in Regional Fisheries Organizations: A Dilemma between Substantive Fishing Interest and Independent Statehood," *Taiwan International Law Quarterly*, Vol. 2, No. 1(2005), pp. 56–57.

[70] *See* the report of the Fourth Meeting of the IOTC, available at *http://www.iotc.org/files/proceedings/1999/s/IOTC-1999-S04-R[EN].pdf* (visited on 03/09/2012).

[71] *See* the report of the Sixth Meeting of the IOTC, available at *http://www.iotc.org/files/proceedings/2001/s/IOTC-2001-S06-R[EN].pdf* (visited on 03/09/2012).

[72] On the resolution, *see* Chapter 3 of this book.

[73] Chiang, *supra note* 69, pp. 57–58.

and FAO in continuing efforts to explore ways on how to effectively manage the fishing fleets of Taiwan Province of China.[74]

After many sessions discussing Taiwan's problem, the IOTC took the huge step of deciding to place itself outside the FAO. This decision was reached in the Ninth Meeting in 2005, at which the IOTC decided to hold a special session to explore how to create a more effective and efficient organisation through changing its relationship with the FAO. For the special session, a draft amendment to the IOTC Agreement and a draft declaration would be prepared for adoption, declaring that amendments to the agreement would not be considered as creating new obligations.[75] It was significant that the IOTC decided that the amendments would not involve new obligations because pursuant to Articles XX(4) and (5) of the IOTC Agreement, an amendment that does not involve new obligations can take effect immediately. Otherwise, the amendment can come into force only after acceptance by each member, which would lead to a complicated relation between the IOTC and its members.[76]

---

[74] *See* the report of the Eighth Meeting of the IOTC, available at *http://www.iotc.org/files/proceedings/2003/s/IOTC-2003-S08-R[EN].pdf* (visited on 05/09/2012).

[75] Details of the IOTC decision concerning the Special Session can be found in the report of the Ninth Meeting of the IOTC, pp. 10–11, available at *http://www.iotc.org/files/proceedings/2005/s/IOTC-2005-S9-R[EN].pdf* (visited on 05/09/2012). Clearly, the IOTC amendment did not favour the PRC's position because it tried to create another route for Taiwan's participation as an independent actor in the Indian Ocean management organisation, instead of maintaining the present situation in which Taiwan was categorised as a province of the PRC. However, there was no record revealing any objection by or attitude of the PRC on this issue.

[76] Article XX(4) of the IOTC Agreement reads: "Amendments not involving new obligations for Members of the Commission shall take effect for all Members from the date of their adoption by the Commission, subject to paragraph 3 above." Article XX(5) reads: "Amendments involving new obligations for Members of the Commission shall, after adoption by the Commission, subject to paragraph 3 above, come into force in respect of each Member only upon its acceptance thereof. The instruments of acceptance of amendments involving new obligations shall be deposited with the Director-General. The Director-General shall inform all Members of the Commission and the Secretary-General of the United Nations of such acceptance. The rights and obligations of any Member of the Commission that has not accepted an amendment involving new obligations shall continue to be governed by the provisions of this Agreement in force prior to the Amendment."

In the Third Special Session in 2006, the IOTC members, in accordance with Article XX of the IOTC Agreement[77], reached consensus on the amendments, including the Proposed Amendments to the IOTC Agreement, Proposed Amendments to the IOTC Rules of Procedure and Draft Amended IOTC Financial Regulations.[78] These would change the IOTC's relationship with the FAO, taking it from within the FAO framework and becoming independence while still cooperating with the FAO. Pursuant to Article XX(2) of the IOTC Agreement, the FAO director-general should have circulated these draft amendments among all members 120 days before the Tenth Meeting in May 2007 for consideration and formal adoption. However, the FAO director-general refused to circulate them, preventing those amendments from being adopted in the next IOTC meeting and making the amendment route was

[77] Article XX of the IOTC Agreement regulates the amendment of the IOTC Agreement. It states:

> "1. This Agreement may be amended by a three-quarters majority of the Members of the Commission.
>
> 2. Proposals for amendments may be made by any Member of the Commission or by the Director-General. Proposals made by a Member of the Commission shall be addressed to both the Chairperson of the Commission and the Director-General and those made by the Director-General shall be addressed to the Chairperson of the Commission, not later than 120 days before the Session of the Commission at which the proposal is to be considered. The Director-General shall immediately inform all Members of the Commission of all proposals for amendments.
>
> 3. Any amendment to this Agreement shall be reported to the Council of FAO which may disallow an amendment which is clearly inconsistent with the objectives and purposes of FAO or the provisions of the Constitution of FAO.
>
> ……
>
> 6. Amendments to the Annexes to this Agreement may be adopted by a two-thirds majority of the Members of the Commission and shall come into force from the date of approval by the Commission.
>
> 7. The Director-General shall inform all Members of the Commission, all Members and Associate Members of FAO and the Secretary-General of the United Nations of the entry into force of any amendment."

[78] The full text of these three amendments can be found in the report of the Third Special Session of the IOTC, Appendices IV, VI and VII, available at *http://www.iotc.org/files/proceedings/2006/s/IOTC-2006-SS3-R[EN].pdf* (visited on 05/09/2012).

unsuccessful. [79] The FAO director-general claimed that it would be inappropriate to directly circulate an amendment to an agreement concluded under Article XIV of the FAO Constitution[80] without seeking

---

[79] William Edeson, "An International Legal Extravaganza in the Indian Ocean: Placing the Indian Ocean Tuna Commission outside the Framework of FAO," *The International Journal of Marine and Coastal Law*, Vol. 22(2007), pp. 487–488.

[80] Article XIV of the FAO Constitution describes the matters of the convention or an agreement relating to the FAO. Paragraph 3 reads:
"Conventions, agreements, and supplementary conventions and agreements shall:
(a) be submitted to the Conference or Council through the Director-General on behalf of a technical meeting or conference comprising Member Nations, which has assisted in drafting the convention or agreement and has suggested that it be submitted to Member Nations concerned for acceptance;
(b) contain provisions concerning the Member Nations of the Organization, and such nonmember States as are members of the United Nations, any of its specialized agencies or the International Atomic Energy Agency, and regional economic integration organizations, including Member Organizations, to which their Member States have transferred competence over matters within the purview of the conventions, agreements, supplementary conventions and agreements, including the power to enter into treaties in respect thereto, which may become parties thereto and the number of acceptances by Member Nations necessary to bring such convention, agreement, supplementary convention or agreement into force, and thus to ensure that it will constitute a real contribution to the achievement of its objectives. In the case of conventions, agreements, supplementary conventions and agreements establishing commissions or committees, participation by non-member States of the Organization that are members of the United Nations, any of its specialized agencies or the International Atomic Energy Agency or by regional economic integration organizations other than Member Organizations, shall in addition be subject to prior approval by at least two-thirds of the membership of such commissions or committees. Where any convention, agreement, supplementary convention or agreement provides that a Member Organization or a regional economic integration organization that is not a Member Organization may become a party thereto, the voting rights to be exercised by such organizations and the other terms of participation shall be defined therein. Any such convention, agreement, supplementary convention or agreement shall, where the Member States of the Organization do not participate in that convention, agreement, supplementary convention or agreement, and where other parties exercise one vote only, provide that the organization shall exercise only one vote in any body established by such convention, agreement, supplementary convention or agreement, but shall enjoy equal rights of participation with Member Nations parties to such convention, agreement, supplementary convention or agreement;
(c) not entail any financial obligations for Member Nations not parties to it other than their contributions to the Organization provided for in Article XVIII, paragraph 2 of this Constitution."

guidance from the Committee on Constitutional and Legal Matters (CCLM) and the FAO Council as to the legally correct course of action to take.[81] The FAO director-general's view could expand the FAO's power over the IOTC, leading to a controversial issue. Article XIV of the FAO Constitution did not endow the director-general with the rights to take this action; however, Article XX of the IOTC Agreement clearly stipulates that the IOTC Agreement could be amended and that the FAO director-general should immediately inform all members of the commission of all proposals for amendments.

At the Eleventh Meeting of the ITOC in 2007, the amendment was discussed,[82] and members expressed their opinions on changing the relationship between the IOTC and the FAO. China was concerned that:

> [G]iven the delay in the distribution of the amendments [it] did not have enough time to properly consider the matter and consequently [was] not able to put forward an opinion to the Commission at this time.[83]

On the other hand, Australia, the European Community, France, Japan, Korea, the Philippines, Sudan and the United Kingdom expressed a positive position on the amendment:

> [T]he change process should go ahead by implementing the changes to the IOTC Agreement as they were proposed in the 3rd Special Session. Inter alia they argued that any binding interpretation of the IOTC Agreement (such as the case that the proposed change could be effected under the current Agreement) can only be made by the IOTC Members; furthermore, the change process is practical, legally acceptable, introduced no new obligations to Members and could be achieved quickly.[84]

At the conclusion of discussions, the members unanimously agreed on the desirability of creating a more effective and efficient commission and invited FAO to take immediate action on the draft amendments in

---

[81] The result was that the FAO Council endorsed the conclusions of the CCLM that the issue was unprecedented and so complex that it was needed to set up an informal group of legal experts to examine the matter. *See* Edeson, *supra note* 79, pp. 490, 492.

[82] *See* the report of the Eleventh Meeting of the IOTC, pp. 6–7available at *http://www.iotc.org/files/proceedings/2007/s/IOTC-2007-S11-R[E].pdf* (visited on 05/11/2012).

[83] *Ibid.*, p. 7.

[84] *Ibid.*

preparation for their further consideration at the IOTC's Twelfth Session.[85] However, this issue so far has never been discussed further.

Although the attempt at amending the IOTC Agreement did not work, it raised many legal problems between the IOTC and the FAO, including the legality of the IOTC amending its own agreement to place itself outside the FAO framework; the legality of the FAO director-general's refusal to circulate the amendments; and the fact if the IOTC has international personality, the legal controversy causes different point of views between two sides, leading to the deadlock.[86] In respect to the IOTC's independence, the FAO suggests that the only way for the IOTC to withdraw from the FAO framework is for the IOTC members to terminate it and then establish a new commission under a new agreement.[87] This procedure is unrealistic because the process of withdrawing from the agreement, terminating the IOTC, adopting a new agreement and establishing a new RFMO would take a long time. During this period, the fishing conducted by each state in the Indian Ocean could cause chaos due to the lack of management, seriously endangering the conservation of fishery resources.

At present, as a temporary expedient, the invited experts from Taiwan may attend IOTC meetings, but the IOTC is still forced to close the door on Taiwan itself. The obstacles to Taiwan's participation show that the IOTC became another battlefield in the conflict between the PRC and Taiwan, sacrificing the real interests of fishery conservation and management.[88]

---

[85] *Ibid.*

[86] William Edeson discusses the related legal problems in depth and presents a detailed consideration of the IOTC and the FAO. *See* William Edeson, *supra note* 79, pp. 485–515; William Edeson, "Some Future Direction for Fishing Entities in Certain Regional Fisheries Management Bodies," *Ocean Development and International Law*, Vol. 37(2006), pp. 251–261.

[87] *See* William Edeson, *supra note* 79, p. 488.

[88] In the latest IOTC Meeting in 2012, to leave an impression that Taiwan is part of the PRC, it continues to declare that all vessels of Taiwan are under the management of China and that China has full control of those vessels, although it does not in reality. *See* the Report of the Sixteenth Meeting of the IOTC, available at *http://www.iotc.org/files/proceedings/2012/s/IOTC-2012-S16-R[E].pdf* (visited on 05/11/2012).

3. Conclusion

Although Taiwan participates in RFMOs under the identity of fishing entity rather than a state, the PRC remains concerned about the nomenclature that Taiwan uses and seeks to avoid an understanding that Taiwan is an independent state. The nomenclature preferred by the PRC is Taiwan, Province of China, as used in the IOTC, because it indicates that Taiwan is part of China. On the other hand, Taiwan prefers its official name of the ROC in the RFMOs, followed by Taiwan, which reflects the reality that Taiwan is a state or at least does not link it to China. The nomenclature mutually acceptable to both the PRC and Taiwan is Chinese Taipei (see Table 8-1). For the PRC, the word "Chinese" emphasises the relation between Taiwan and China, and the word "Taipei" refers merely to one city in Taiwan, enabling Chinese Taipei to be regarded as a territory or region belonging to China. For Taiwan, the term "Chinese" refers to Chinese culture, including Mandarin speakers, which Taiwan claims as its own culture, and to Taipei, the capital of Taiwan. Thus, "Chinese Taipei" indicates Taiwan's capital, Taipei, and Chinese culture. In RFMOs in which the PRC participates, Taiwan can only use the nomenclature "Chinese Taipei."

The nomenclature of Taiwan in RFMOs is a petty matter but reflects the differential treatment of Taiwan in treaties to which the PRC is not a party. Under the CCSBT in which the PRC is not a contracting party or member, Taiwan can participate under the name Fishing Entity of Taiwan, which is closer to Taiwan's second choice and does not imply that Taiwan is part of or under the control of China. In the case of the ICCAT, the PRC government replaced the Taiwanese government in its UN seat in 1971, but Taiwan still attended the 1972 ICCAT meetings as an observer with the identity of a state and the name Taiwan in 1972. At that time, the PRC was not a party of the ICCAT Convention, and the ICCAT never challenged or opposed Taiwan's status as a state. However, when the PRC became a party of the ICCAT Convention in 1996, it protested the ICCAT allowing Taiwan to participate under the name Taiwan and that some ICCAT documents categorised Taiwan as a "non contracting nation." Although the ICCAT explained that its practice did not imply any recognition, Taiwan was forced to change its name to Chinese Taipei.

Table 8-1: Taiwan's Participation in RFMOs

RFMOs	Date of Participation	Status	Identity	Nomencla -ture
CCSBT	20 May 1994	Observer	Entity	Taiwan Observer
CCSBT Extended Commission	30 Aug. 2002	Member	Fishing entity	Fishing Entity of Taiwan
IATTC	27 Aug. 2010	Member	Fishing entity	Chinese Taipei
ICCAT	1972	Observer	State	Fisheries of Taiwan
	2001	Cooperat -ing fishing entity	Fishing entity	Chinese Taipei
IOTC	2001	Invited expert	Individuals	Taiwan, Province of China
ISC	29 Jan. 2002	Member	Fishing entity	Chinese Taipei
WCPFC	19 Jun. 2004	Member	Fishing Entity	Chinese Taipei

The diplomatic competition between the ROC and the PRC affects Taiwan's participation in RFMOs, both those in which it is a member and those in which it cannot acquire the status of a member. Comparing Taiwan's participation as a non-member in various RFMOs (see Table 8-1), Taiwan enjoys the best status, that of a quasi-member, in the CCSBT, which demonstrated its flexibility in effectively bringing Taiwan into its regime by establishing the Extended Commission. Doing so was possible mainly because fewer states are involved, making the situation less complicated and reducing the PRC's influence on the commission. Although Taiwan is only an observer of the CCSBT, it is nearly fully involved in the whole operation of the CCSBT through its member status in the Extended Commission. With respect to the ICCAT, Taiwan could not obtain the status of a member, so it sought to normalise its status as a cooperating fishing entity, which was an improvement over the unstable status of an observer and seems to be the best solution for participation in the ICCAT. At present, the IOTC is the only tuna RFMO that has not found a solution to enable Taiwan's participation. Being within the FAO/UN framework which supports the PRC's One China Policy leads to inflexibility and poses the main obstacle for both the IOTC and Taiwan. As far as the IOTC is concerned, bringing Taiwan into its regime would greatly benefit the effectiveness of its conservation and management because Taiwan is a major catcher in the Indian Ocean. Hence, the IOTC tried to place itself outside the FAO framework by amending the IOTC Agreement, causing a legal controversy with the FAO. Thus, Taiwan attends IOTC meetings as an observer under the identity of invited experts, rather than a fishing entity. The IOTC's attempt to placing itself outside the FAO reflects the necessity of Taiwan's participation. Neither the IOTC nor the FAO should be willing to see the states under the IOTC regime burdened with the cost caused by the nonparticipation of the Taiwanese government. Hence, it is expected that a compromise will be found for Taiwan's participation in the future.

CONCLUSION

Due to the political conflict with the PRC, Taiwan is not recognised as a state by most countries and so faces obstacles to participate in RFMOs under the identity of a state. However, to effectively carry out conservation and management regulations, regimes need the cooperation of Taiwan, with its high capability for fishing. In addition, Taiwan seeks to promote its international image as a good actor willing to cooperate with other states in order to benefit its competition with the PRC. Consequently, to participate in RFMOs, Taiwan compromised on its original intention to use the identity of a fishing entity through the focus on its function of fishing. Taiwan regards itself as a state but adopts the identity of fishing entity only in RFMO regimes. Notwithstanding, a position for fishing entities different than that of states has been created in international fishery regulations.[1]

The 1995 UNFSA creates a door for fishing entities to participate in conservation and management regimes for straddling and highly migratory fish stocks through individual RFMOs. The lack of a definition of the term "fishing entity" leads to the absence of standard criteria for the elements which constitute a fishing entity. Research question (a) asked why the issue of fishing entities is important and what the legal status of fishing entities is under the international law of the sea and RFMOs. Fishing entities are different than states, which can be bound not only by RFMOs but also by the UNFSA and UNCLOS. Without a route for fishing entities to become parties to the UNCLOS and the UNFSA, RFMOs are the only agencies that can prevent fishing entities from violating conservation and management measures. However, if fishing entities are not willing or able to join the RFMOs, they might stay outside conservation and management regimes. This would further reduce the effectiveness of the multi-protection net created by the UNCLOS, UNFSA and RFMOs, especially when a fishing entity plays a major role in fishing in many high sea areas. Thus, the issue of fishing entities is important.

Although fishing entities face difficulty becoming parties to the UNCLOS and the UNFSA, it cannot be denied that they possess international legal

[1] Through viewing Taiwan as an illustration, a fishing entity can be defined as an actor or entity which possesses the same legal capability in fishing as a state.

personality within regimes in which they are equivalent to states as the subject of regulations.[2] It means that a fishing entity, although other than a state, can still be the subject of the agreements which create RFMOs. However, this gives rise to the problems that a fishing entity or international community might face under present fishery regulations, *i.e.* research question (c). Firstly, with respect to the conservation and management of straddling and highly migratory fish stocks, the UNCLOS, UNFSA and RFMOs form a multi-protection net, in which UNCLOS is the base, the UNFSA is the core and RFMOs are the units of implementation. States are inside the regime and required to be bound by these three systems of regulations. Of course, it is possible that a state is a party to only one or two of these instruments; however, the state is not excluded from these three treaties but still could be bound by all treaty systems. Due to the inability to become a party to the UNCLOS and UNFSA, fishing entities exist only at the RFMO level. Notwithstanding the principle of *pacta tertiis nec nocent nec prosunt*, if a fishing entity joins in the RFMOs, it might indirectly comply with the related UNCLOS and the UNFSA regulations. If it refuses to be involved in RFMOs, then there is no other international agreement that can force it to follow the related rules. In other words, RFMOs become the crucial element bringing fishing entities into the system; thus, a fishing entity can reduce the multi-protection function of the triangle system to a single layer of protection. This situation does not benefit the effectiveness of the conservation and management regime because it depends completely on the fishing entity's willingness to comply. The fishing entity could be completely outside the regime if it refuses to participate in any organisation. This also influences the effectiveness of high seas enforcement measures by the UNFSA or RFMOs.

Furthermore, Article 21(2) of the UNFSA stipulates that states should through RFMOs establish a boarding and inspection procedure. Article 21(3) of the UNFSA states that, if a RFMO does not establish its own boarding and inspection procedure by two years after the adoption of the UNFSA, then the UNFSA Boarding and Inspection Procedure in Article 22 will apply automatically as between UNFSA parties. Under Article 21 of the UNFSA, the RFMOs are expected to establish their own boarding and inspection procedures. However, this article does not indicate whether the RFMOs to which it refers are those set up after the 1995 UNFSA or those that existed before 1995. If it refers to those established after 1995, then the UNFSA still needs to create a link to RFMOs which existed

[2] *See* Chapter 2 of this book.

before because they have no obligation to cooperate with the UNFSA. Even RFMOs established after 1995 are not obliged to comply with the UNFSA. This uncertain relation between the UNFSA and RFMOs can cause some difficulties in implementing boarding and inspection procedure regulations within the UNFSA. In this situation, if a state is a member of a RFMO and a state party of the UNFSA, it might be simpler to implement the boarding and inspection procedure between individual members. However, if a fishing entity can only become a member of a RFMO, then the problem is whether Article 22 of the UNFSA can apply to it. If the RFMO decides to comply with Article 21 of the UNFSA, then this fishing entity is indirectly bound by Articles 21 and 22. However, if a fishing entity is not willing to become a member of a RFMO which complies with Article 21 of the UNFSA, then a loophole allows this fishing entity to escape high seas enforcement.

As RFMOs are the main means for a fishing entity to participate in conservation and management regimes, the degree of a fishing entity's engagement in individual RFMOs is worth assessing. Among the RFMOs discussed, a fishing entity could obtain the best status in the ISC because it distinguishes between members and non-members, instead of between contracting and non-contracting parties. A state and a fishing entity could join the ISC as members if they meet the qualifications in the ISC Charter. The fishing entity would have exactly the same status as other state members. The IATTC provides the second best status for fishing entities. Article 1(7) of the Antigua Convention clearly stipulates that members include contracting parties and fishing entities. Although a fishing entity may not become a contracting party of the Antigua Convention, it is guaranteed to enjoy the same rights and to discharge the same obligations as other state members through the convention's *mutatis mutandis* approach. In addition, a fishing entity can be bound by the Antigua Convention through a signature or a commitment of an instrument, establishing clearer legal relations among the fishing entity, convention and other contracting parties.

Fishing entities' status in the WCPFC is similar to in the IATTC. A fishing entity cannot become a contracting party but only be a member through the delivery of an instrument which is stipulated in the Annex of the WCPFC Convention, instead of the body text. As a member, the fishing entity has mostly the same rights as contracting parties, except for the eligibility to be elected as chairman and vice-chairman, decide the location of headquarters of the commission and appoint the executive secretary. The most interesting RFMO regarding the status of fishing entity is the CCSBT. To solve the difficulty of fishing entities' participation and avoid the delay that might be caused by amending the CCSBT Convention, it established an Extended Commission, comprised of all existing CCSBT members and any fishing entity whose vessels conduct fishing activities for SBT in the CCSBT convention area. The decision making process was essentially transferred from the original CCSBT Commission to the Extended Commission, in which the fishing entity has the same position as other state members. The CCSBT Commission could refuse to adopt the Extended Commission's decisions, preventing them from being implemented. However, the membership of the two bodies is nearly identical, so it seems unlikely that the same members would reject a decision they had made. Thus, the Extended Commission has essentially taken over the function of the CCSBT Commission. A fishing entity may not be a contracting party or even a full member of the CCSBT but, through the Extended Commission, can fully participate in and influence the decisions of the CCSBT.

The clever design of the Extended Commission in the CCSBT is worth considering as an approach to the same difficulty of fishing entities' full participation in the ICCAT and IOTC. However, these two RFMOs have many more members than the CCSBT, making it more difficult and complicated to create an extended commission of which all members approve and agree to join. In the ICCAT, a fishing entity cannot be a contracting party nor a member but only a cooperating fishing entity which does not enjoy the same rights as other members but discharges the same obligations. Although the ICCAT allows a fishing entity to participate in allocation of fishing quota, it has a status far lower than that of other members. The RFMO which assigns the worst position to fishing entities is the IOTC, which is inside the UN framework; hence, its membership is restricted to UN members and specialised agencies. The only way for a fishing entity to participate in the IOTC is to apply as a cooperating non-contracting party, which obliges the fishing entity to cooperate with the IOTC's conversation and management measures and allows it to fish in the IOTC Convention area. However, the critical problem that the IOTC faces concerning fishing entities is posed by Taiwan, which is the major catcher of tuna in the Indian Ocean but excluded from the IOTC mainly due to its political dispute with China. Consequently, the effectiveness of the IOTC's conservation and management measures is sacrificed for political reasons.

Research question (b) asked whether fishing entities' status in fisheries enforcement on the high seas is to that of a flag state or a non-flag state. Analysis of the UNFSA and RFMO regulations shows that a fishing entity can enjoy the freedom of fishing on the high seas. A fishing entity does not face barriers to acting in high seas enforcement against vessels as the equivalent of a flag state or to acting as the equivalent of a non-flag state in high seas enforcement through bilateral treaties with other entities or states. In other words, a fishing entity conducts high seas enforcement against its vessels as a flag state on the basis of its responsibility to ensure that its vessels do not violate rules. As a non-flag state, a fishing entity can conclude bilateral treaties to conduct enforcement against other states' vessels. Theoretically, a fishing entity can also do so through multilateral treaties, including through RFMOs. However, due to their different status compared to other state members and contracting parties, fishing entities' position in high seas enforcement within RFMOs is more complicated. No matter whether at the bilateral or regional treaty level, non-flag state enforcement on the high seas by fishing entities is limited to boarding and inspection; further actions such as investigation or prosecution remain the jurisdiction of the flag state.

The WCPFC is the only RFMO which has adopted a boarding and inspection procedure involving fishing entities since the adoption of the UNFSA. Under the procedure, contracting parties of the WCPFC Convention can board and inspect each other's vessels without any expression of their agreement because they have an obligation to conform to the procedure. However, as a member of the WCPFC, a fishing entity must deliver notification expressing its willingness to comply with the Boarding and Inspection Procedure, and if all other members express their willingness to apply the procedure with that fishing entity, then the rights and obligations stipulated by the procedure are established between the state members and the fishing entity. This process can lead to a problematic consequence: If a fishing entity is unwilling to comply with the procedure or if it delivers a notification but all or some state members do not express willingness to apply the procedure with that fishing entity, then the fishing entity can be legally outside the WCPFC Boarding and Inspection Regime. This measure not only returns the relation between the fishing entities and contracting parties in WCPFC boarding and inspection regime to the level of bilateral treaties but also creates a loophole for both fishing entities and state members. This situation is also linked to the research question (c).

Regarding the practices for enforcement by fishing entities on the high seas within the RFMOs to which research question (d) refers, only the WCPFC has established its own boarding and inspection procedure and has involved a fishing entity, namely Taiwan. Taiwan has applied the WCPFC Boarding and Inspection Procedure through notification with New Zealand, the Cook Islands, the United States, Japan, France and

Australia. Taiwan has had 13 registered inspection vessels conducting high seas boarding and inspection since 14 July 2008 but has found no violations so far.

Due to the vulnerability of straddling and highly migratory stocks, states cooperate directly or through RFMOs to conserve and manage these fishery resources and protect them from commercial extinction or becoming endangered as species. The importance of these efforts has surpassed that of political conflicts between states because the growth and decline in one species affects other species. The commercial extinction of one or more stocks could cause other species to become endangered and lead to an ecological disaster. Therefore, RFMOs should encourage fishing actors, including fishing entities, to participate in and comply with related regulations in order to increase the possibility of effectively solving the problem of fishery resources. Excluding these fishing actors by leaving them outside conservation and management regimes worsens and complicates the problem. In other words, if the common concern of states is to conserve and manage fish stocks, then RFMOs should be focused on their functional purpose, which is not to resolve political conflicts among states but to bring all actors fishing within their convention area under regulations in order to ensure that all actors comply with the conservation and management measures and fully participate in the organisations' sustainable development schemes and policies.

The best way to prevent excluding fishing entities from the conservation and management regime is to amend the UNCLOS and UNFSA to allow fishing entities to become parties. Thus, RFMOs would not bear the crucial responsibility of deciding whether fishing entities are inside or outside the conservation and management regime. However, it can be presumed that amending these two international agreements would be difficult, complicated and time-consuming, requiring the agreement and ratification of many states. However, this circumstance suggests that new relevant international agreements should allow fishing entities to be parties. If the UNCLOS and UNFSA are not amended, a practical way to encourage fishing entities to participate in the conservation and management regime is to grant them the same status as other member states within RFMOs.

The best model is provided by the ISC, which creates no unnecessarily complicated disparities between members, non-members, contracting parties and non-contracting parties but only between states and fishing entities that are members or non-members. This simpler design is fair to each participant, including fishing entities. If a fishing entity as a member has the same obligations as other member states, it should also enjoy the same rights. This scheme indirectly encourages fishing entities to participate, which increases the effectiveness of conservation and management measures. This model could also be used to resolve the problems regarding the WCPFC Boarding and Inspection Procedure. If the procedure were amended to give the same status to contracting parties and members, then notifications would not be needed for the boarding and inspection procedure between fishing entities and other state members, but the WCPFC procedure itself could be applied directly. High seas fisheries enforcement by fishing entities thus would move from the bilateral to the multilateral level. This change would not only make the conservation and management regime more effective but also set a positive model for other RFMOs in high seas fisheries enforcement by fishing entities.

REFERENCES

Bibliography

Anderson, David H. "The Straddling Stocks Agreement of 1995—An Initial Assessment." *International and Comparative Law Quarterly* 45 (1996): 463–475.

"Asia: Practising to Provoke; Counter-proliferation." *The Economist* 368 (Sept. 20, 2003): 79.

Balton, David A. "Strengthening the Law of the Sea: The New Agreement on Straddling Fish Stocks and Highly Migratory Fish Stocks." *Ocean Development & International Law* 27 (1996): 125–151.

Black, Henry Campbell. *Black's Law Dictionary*, 5th ed. St. Paul, Minn.: West Publishing Co., 1979.

Brown, E.D. *The International Law of the Sea*, vol. 1. London: Dartmouth Publishing Company, 1994.

Brownlie, Ian. *Principles of Public International Law*, 6th ed. New York: Oxford University Press, 2003.

—. *Principles of Public International Law*, 7th ed. New York: Oxford University Press, 2008.

Byers, Michael. "Policing the High Sea: The Proliferation Security Initiative." *American Journal of International Law* 98, no.3 (July, 2004): 526–545.

Cheng, Bin. "Introduction to Subjects of International Law." In *International Law: Achievements and Prospects*, edited by Mohammed Bedjaoui, 23-40. Dordrecht: Martinus Nijhoff Publishers, 1991.

Chiang, Huang-Chih. "On Taiwan's Participation in Regional Fisheries Organizations: A Dilemma between Substantive Fishing Interest and Independent Statehood." *Taiwan International Law Quarterly* 2, no. 1 (2005): 17–93.

Churchill, R.R. and A.V. Lowe. *The Law of the Sea*, 3rd ed. Manchester: Manchester University Press, 1999.

Colombos, C.J. *The International Law of the Sea*, 6th ed. London: Longmans Green & Co. Ltd., 1967.

Davies, Peter G.G. and Catherine Redgwell. "The International Legal Regulation of Straddling Fish Stocks." *The British Yearbook of International Law* 67 (1996): 199–274.

Djalal, Hasjim. "The Emergence of the Concept of Fishing Entities: A Note." *Ocean Development and International Law* 37 (2006): 117–121.

Dupuy, René-Jean and Daniel Vignes, eds. *A Handbook on the New Law of the Sea*. Netherlands: Martinus Nijhoff Publishers, 1991.

Ebbin, Syma A. *A Sea Change: The Exclusive Economic Zone and Governance Institutions for Living Marine Resources*. Dordrecht: Springer, 2005.

Edeson, W.R. "An International Legal Extravaganza in the Indian Ocean: Placing the Indian Ocean Tuna Commission outside the Framework of the FAO." *The International Journal of Marine and Coastal Law* 22, no.4 (2007): 485–515.

Edeson, William. "Some Future Directions for Fishing Entities in Certain Regional Fisheries Management Bodies." *Ocean Development and International Law* 37 (2006): 245–264.

Evans, Malcolm D. *International Law Documents*, 8th ed. Oxford: Oxford University Press, 2007.

Fenby, Jonathan. *Generalissimo: Chiang Kai-Shek and the China He Lost*. London: Free Press, 2005.

Food and Agriculture Organization of the United Nations. *FAO Yearbook: Fishery and Aquaculture Statistics*. Rome: FAO Fisheries Department, 2009.

Food and Agriculture Organization of the United Nations. *FAO Yearbook: Fishery and Aquaculture Statistics*. Rome: FAO Fisheries Department, 2012.

Graham, Robert A. *Vatican Diplomacy*. Princeton: Princeton University Press, 1959.

Grotius, Hugo. *The Freedom of the Seas*. Translated by Ralph Van Deman Magoffin. New York: Oxford University Press, 1916.

Guilfoyle, Douglas. *Shipping Interdiction and the Law of the Sea*. New York: Cambridge University Press, 2009.

Harris, D.J. *Cases and Materials on International Law*, 6th ed. London: Sweet & Maxwell, 2004.

—. *Cases and Materials on International Law*, 7th ed. London: Sweet & Maxwell, 2010.

Ho, Peter S.C. "The Impact of the U.N. Fish Stocks Agreement on Taiwan's Participation in International Fisheries fora." *Ocean Development and International Law* 37 (2006): 133–148.

Hyde, Charles Cheney. *International Law: Chiefly as Interpreted and Applied by the United States*, vol. 1, 2nd rev., ed. Boston: Little, Brown and Company, 1951.

Hu, Nien-Tsu Alfred. "Fishing Entities: Their Emergence, Evolution, and Practice from Taiwan's Perspective." *Ocean Development and International Law* 37 (2006): 149–183.

International Law Commission. *Yearbook of the International Law Commission*, vol. 1. New York: United Nations, 1951.

Johnston, Douglas M. *Marine Policy and the Coastal Community: The Impact of the Law of the Sea*. London: Croom Helm Ltd., 1976.

Lévy, Jean-Pierre and Gunnar G. Schram, eds. *United Nations Conference on Straddling Fish Stocks and Highly Migratory Fish Stocks—Selected Documents*. Hague: Kluwer Law International, 1996.

Lodge, Michael W. "The Practice of Fishing Entities in Regional Fisheries Management Organizations: The Case of the Commission for the Conservation and Management of Highly Migratory Fish Stocks in the Western and Central Pacific Ocean." *Ocean Development and International Law* 37 (2006): 185–207.

Lynden, D. W. "The Convention on Conduct of Fishing Operations in the North Atlantic." *Netherlands International Law Review* 14 (1967): 245–258.

McDougal, Myres S. and William T. Burke. *The Public Order of the Oceans: A Contemporary International Law of the Sea*. New Haven: Yale University Press, 1962.

Moore, John Bassett. *A Digest of International Law*, vol. 2. Mich.: UMI, A Bell & Howell Company, 1993.

Netherlands Institute for the Law of the Sea. *International Organizations and the Law of the Sea: Documentary Yearbook*, vol. 1. Alphen aan den Rijn: Kluwer Law International, 1985.

O'Connell, D.P. *International Law*, vol. 1, 2nd ed. London: Stevens and Sons, 1970.

Oppenheim, L. *Oppenheim's International Law*, vol. I, 9th ed., edited by Robert Jennings and Arthur Watts. London: Longman, 1992.

Orrego, Vicuña F. *The Changing International Law of High Seas Fisheries*. New York: Cambridge University Press, 1999.

Potter, Pitman B. *The Freedom of the Seas, in History, Law and Politics*. London: Longmans, Green & Co., 1924.

Rayfuse, Rosemary Gail. *Non-flag State Enforcement in High Seas Fisheries*. Leiden: Martinus Nijhoff Publishers, 2004.

Rayfuse, Rosemary. "Regulation and Enforcement in the Law of the Sea: Emerging Assertions of a Right to Non-flag State Enforcement in the High Seas Fisheries and Disarmament Contexts." *Australian Year Book of International Law* 24 (August 2005): 181–200.

—. "The Interrelationship between the Global Instruments of International Fisheries Law." In *Developments in International Fisheries Law*, edited by Ellen Hey, 107-158. Hague: Kluwer Law International, 1999.

Reuland, Robert C.F. "Interference with Non-national Ships on the High Seas: Peacetime Exceptions to the Exclusivity of Flag State Jurisdiction." *Vanderbilt Journal of Transnational Law* 22 (1989): 1161–1229.

Roy, Denny. *Taiwan: A Political History*. New York: Cornell University Press, 2003.

Serdy, Andrew. "Bringing Taiwan into the International Fisheries Fold: the Legal Personality of a Fishing Entity." In *The British Year Book of International Law 2004*, vol. 75, edited by James Crawford and Vaughan Lowe 183–221. Oxford: Clarendon Press, 2005.

Serdy, Andrew. "Postmodern International Fisheries Law, or We Are All Coastal States Now." *The International and Comparative Law Quarterly* 60 (April 2011): 387–422.

Smith, H.A. *The Law and Custom of the Sea*, 3rd ed. London: Stevens and Sons Limited, 1959.

Tsamenyi, Martin. "The Legal Substance and Status of Fishing Entities in International Law: A Note." *Ocean Development and International Law* 37 (2006): 123–131.

Wang, Dustin Kuan-Hsiung. "Taiwan's Participation in Regional Fisheries Management Organizations and the Conceptual Revolution on Fishing Entity: The Case of the IATTC." *Ocean Development and International Law* 37 (2006): 209–219.

International Agreements and Documents

Treaties

"International Convention for Regulating the Police of the North Sea Fisheries," 1882, <http://iea.uoregon.edu/pages/view_treaty.php?t=1882-PoliceNorthSeasFishery.EN.txt&par=view_treaty_html> (accessed December 13, 2011).

"Convention Respecting the Liquor Traffic in the North Sea," 1887, <http://www.legislation.gov.uk/ukpga/Vict/56-57/17/schedule> (accessed May 27, 2013).

"Treaty of Shimonoseki," 1895, <http://china.usc.edu/(S(u0o4uo4552nc4b45rm1khibm)A(4BUbwWp7zQEkAAAAZDlkMDBjYTctYjM1NC00ODg1LWE5YjgtYTg4NGE0

ZDc0ODU4BJSk2ve3v12T5hSKPbdqc1b-n8A1))/ShowArticle.aspx?articleID=405&AspxAutoDetectCookieSupport=1> (accessed June 1, 2012).

"Statute of the International Court of Justice," 1945, <http://www.icj-cij.org/documents/index.php?p1=4&p2=2&p3=0> (accessed May 3, 2011).

"International Convention for the High Seas Fisheries of the North Pacific Ocean," 1952, <http://sedac.ciesin.columbia.edu/entri/texts/fisheries.north.pacific.1952.html> (accessed December 19, 2012).

"International Convention for the Unification of Certain Rules Relating to Penal Jurisdiction in Matters of Collision or Other Incidents of Navigation," 10 May 1952, United Nations, *Treaty Series*, vol. 439, 235–241. New York: United Nations, 1962.

"Convention on the High Seas," 29 April 1958, United Nations, *Treaty Series,* vol. 450, 11. New York: United Nations, 1964.

"International Convention for the Conservation of Atlantic Tunas," 1966, <http://www.iccat.int/Documents/Commission/BasicTexts.pdf> (accessed December 21, 2010).

"Convention on Conduct of Fishing Operations in the North Atlantic," 1967, <http://www.fco.gov.uk/resources/en/pdf/3706546/3892723/21060066/Tr-FishNorthAtlan-Ts40.1977> (accessed January 3, 2012).

"Convention on Fishing and Conservation of the Living Resources of the High Seas," United Nations, *Treaty Series*, vol. 559, 285–292. New York: United Nations, 1967.

"Agreement between the Government of the United States of America and the Government of the People's Republic of Bulgaria Concerning Fisheries off the Coasts of the United States," 1976, United Nations, *Treaty Series,* vol. 1134, 127.

"Agreement between the Government of the United States of America and the Government of the Polish People's Republic Concerning Fisheries off the Coasts of the United States," 1976, <http://untreaty.un.org/unts/144078_158780/3/5/11639.pdf> (accessed December 11, 2011).

"Agreement between the Government of the United States of America and the Government of the Republic of China Concerning Fisheries off the Coasts of the United States," 1976, <http://www.intfish.net/treaties/bilaterals/texts/chi-usa/chi-usa-1976.pdf> (accessed November 28, 2007).

"Agreement between the Government of the United States of America and the Government of the Socialist Republic of Romania Concerning Fisheries off the Coasts of the United States," 1976, <http://www.intfish.net/treaties/bilaterals/texts/rom-usa/rom-usa-1976.pdf> (accessed November 28, 2007).

"Agreement between the Government of the United States of America and the Government of the Union of Soviet Socialist Republics on Mutual Fisheries Relations," 1976, <http://www.state.gov/documents/organization/138873.pdf> (accessed January 4, 2012).

"Agreement between the Government of Spain and the Government of the United States of America Concerning Fisheries off the Coasts of the United States," 1977, <http://www.intfish.net/treaties/bilaterals/texts/spa-usa/spa-usa-1977.pdf> (accessed November 28, 2007).

"Agreement between the Government of the United States of America and the Government of Mexico Concerning Fisheries off the Coasts of the United States," 1977, <http://www.intfish.net/treaties/bilaterals/texts/mex-usa/mex-usa-1977.pdf> (accessed November 28, 2007).

"Agreement between the Government of the United States of America and the Government of the Republic of Cuba Concerning Fisheries off the Coasts of the United States," 1977, in "Cuba—United States: Agreement Concerning Fisheries off the Coasts of the United States," *International Legal Materials*, 16, no. 3 (May 1977): 596–605.

"Convention on Future Multilateral Cooperation in the Northwest Atlantic Fisheries," 1978, <http://www.nafo.int/about/frames/about.html> (accessed January 5, 2012).

"Agreement between the Government of the United States of America and the Government of Portugal Concerning Fisheries off the Coasts of the United States," 1980, <http://faolex.fao.org/docs/texts/por22901.doc> (accessed December 12, 2012).

"Agreement between the Government of the United States of America and the Government of Norway Concerning Fisheries off the Coasts of the United States," 1981, <http://untreaty.un.org/unts/60001_120000/9/21/00017008.pdf> (accessed December 12, 2012).

"Agreement between the Government of the United States of America and the Government of the Republic of Korea Concerning Fisheries off the Coasts of the United States," 1982,

<http://www.intfish.net/treaties/bilaterals/texts/rok-usa/rok-usa-1982a.pdf> (accessed November 28, 2007).

"United Nations Convention on the Law of the Sea," 1982, *A/CONF.62/122*, 1833 UNTS 397.

"Agreement between the Government of the United States of America of the One Part and the Government of Denmark and the Home Government of the Faroe Islands of the Other Part Concerning Fisheries off the Coasts of the United States," 1984, <http://untreaty.un.org/unts/144078_158780/1/1/100.pdf> (accessed December 12, 2011).

"Agreement between the Government of the United States of America and the European Economic Community Concerning Fisheries off the Coasts of the United States," 1984, <http://faolex.fao.org/docs/texts/bi-11049.doc> (accessed December 12, 2011).

"Agreement between the Government of the United States of America and the Government of the Republic of Iceland Concerning Fisheries off the Coasts of the United States," 1984, <http://www.intfish.net/treaties/bilaterals/texts/ice-usa/ice-usa-1984.pdf> (accessed November 28, 2007).

"United Nations Convention on Conditions for Registration of Ships," 1986, <http://www.admiraltylawguide.com/conven/registration1986.html> (accessed November 4, 2011).

"Convention for the Conservation Anadromous Stocks in the North Pacific Ocean," February 1992, <http://www.npafc.org/new/publications/HandBook/English%20(page 1-44).pdf> (accessed January 7, 2011).

"Agreement for the Establishment of the Indian Ocean Tuna Commission," FAO Council at its 105th Session, 25 November 1993, <http://www.iotc.org/files/proceedings/misc/ComReportsTexts/IOTC %20Agreement.pdf> (accessed February 18, 2011).

"Agreement to Promote Compliance with International Conservation and Management Measures by Fishing Vessels on the High Seas," FAO Conference Resolution *15/93* at the 27th Session of the FAO Conference, November 1993, <http://www.fao.org/DOCREP/MEETING/003/X3130m/X3130E00.HTM> (accessed February 15, 2011).

"Convention for the Conservation of Southern Bluefin Tuna," May 1993, <http://www.ccsbt.org/docs/pdf/about_the_commission/convention.pdf> (accessed December 15, 2010).

"Convention on the Conservation and Management of the Pollock Resources in the Central Bering Sea," 16 June 1994, <http://www.afsc.noaa.gov/REFM/CBS/Docs/Convention%20on%20 Conservation%20of%20Pollock%20in%20Central%20Bering%20Sea. pdf> (accessed January 5, 2012).

"Agreement for the Implementation of the Provisions of the United Nations Convention on the Law of the Sea of 10 December 1982 Relating to the Conservation and Management of Straddling Fish Stocks and Highly Migratory Fish Stocks," *A/CONF.164/37*, 2167 UNTS 3, 1995.

"The Grains Trade Convention," 1995, <http://ec.europa.eu/world/agreements/downloadFile.do?fullText=yes &treatyTransId=13582> (accessed Auagust 15, 2012).

"The Food Aid Convention," 1999, <http://www.foodaidconvention.org/Pdf/convention/iga1995.pdf> (accessed Auagust 15, 2012).

"Convention on the Conservation and Management of Highly Migratory Fish Stocks in the Western and Central Pacific Ocean," 5 September 2000, <http://www.wcpfc.int/doc/convention-conservation-and-management-highly-migratory-fish-stocks-western-and-central-pacific-> (accessed August 22, 2010).

"Convention on the Conservation and Management of Fisheries Resources in the South East Atlantic Ocean," April 2001, <http://www.seafo.org/welcome.htm> (accessed January 11, 2011).

"Convention for the Strengthening of the Inter-American Tropical Tuna Commission Established by the 1949 Convention between the United States of America and the Republic of Costa Rica," June 2003, <http://www.iattc.org/PDFFiles2/Antigua_Convention_Jun_2003.pdf> (accessed August 05, 2010).

"The Fishery Agreement between the Association of East Asian Relations and the Interchange Association," (Translated), 10 April 2013, <http://no06.mofa.gov.tw/mofatreatys/ShowPicOut.aspx?FileFolder=0 0&FileName=002013/04/10_C.pdf> (accessed April 04, 2014).

Other International Documents

"Agenda 21," United Nations Conference on Environment & Development, 3–14 June 1992, <http://www.un.org/esa/sustdev/documents/agenda21/english/Agenda2 1.pdf> (accessed February 15, 2011).

"Arrangement for the Participation of Fishing Entities," WCPFC, 4 April 2009, <http://www.wcpfc.int/doc/arrangement-participation-fishing-entities> (accessed August 22, 2010).

"Charter of the International Scientific Committee for Tuna and Tuna-like Species in the North Pacific Ocean," Amended at the Third Plenary Meeting on 29 January 2002 and at the Seventh Plenary Meeting on 25–30 July 2007, <http://isc.ac.affrc.go.jp/about_isc/charter.html> (accessed August 05, 2010).

"Code of Conduct for Responsible Fisheries," FAO Conference, 1995, <ftp://ftp.fao.org/docrep/fao/005/v9878e/v9878e00.pdf> (accessed February 22, 2010).

"Draft Agreement for the Implementation of the Provisions of the United Nations Convention on the Law of the Sea of 10 December 1982 Relating to the Conservation and Management of Straddling Fish Stocks and Highly Migratory Fish Stocks," *A/CONF. 164/22*, 1994, <http://daccess-dds-ny.un.org/doc/UNDOC/GEN/N94/338/70/IMG/N9433870.pdf?OpenElement> (accessed February 17, 2011).

"Draft Agreement for the Implementation of the Provisions of the United Nations Convention on the Law of the Sea of 10 December 1982 Relating to the Conservation and Management of Straddling Fish Stocks and Highly Migratory Fish Stocks," *A/CONF. 164/22/Rev.1*, 1995, <http://daccess-dds-ny.un.org/doc/UNDOC/GEN/N95/105/06/IMG/N9510506.pdf?OpenElement> (accessed February 21, 2011).

"Draft Amended IOTC Financial Regulations," <http://www.iotc.org/files/proceedings/2006/s/IOTC-2006-SS3-R[EN].pdf> (accessed September 05, 2012).

"Draft Procedures for Boarding and Inspection Pursuant to the WCPF Convention," Working III of the Fourth Session of the WCPFC Preparatory Conference, <http://www.wcpfc.int/doc/wcpfcprepconwp14/draft-procedures-boarding-and-inspection-pursuant-convention> (accessed August 10, 2012).

François, J.P.A., *Report on the Regime of the High Seas*, U.N. Doc. No. A/CN.4/17, 1950, <http://untreaty.un.org/ilc/documentation/english/a_cn4_17.pdf> (accessed December 13, 2011).

"Fisheries Report No. 484 Supplement," *FIPL/R484(Suppl.)*, 7–15 September 1992, *Papers Presented at the FAO Technical Consultation on High Seas Fishing*, 44–54.

Gidel, Gilbert, "Memorandum," United Nations Secretariat, *Memorandum on the Regime of the High Seas, U.N. Doc. No. A/CN.4/32*,1950,

<http://untreaty.un.org/ilc/documentation/english/a_cn4_32.pdf> (accessed November 10, 2011).

"Global Southern Bluefin Tuna Catch by Flag," CCSBT, <http://www.ccsbt.org/site/sbt_data.php> (accessed August 15, 2012).

"High Seas Boarding and Inspection Procedure (Revised)," *WCPFC-TCC3-2007-11*, Third Regular Session of WCPFC Technical and Compliance Committee, <http://www.wcpfc.int/doc/wcpfc-tcc3-2007-11/high-seas-boarding-and-inspection-procedures-rev-1> (accessed April 05, 2012).

"ICCAT Scheme of Joint International Inspection," *Report of the Fourth Regular Meeting of the ICCAT*, 1995, <http://www.iccat.int/Documents/BienRep/REP_EN_74-75_II.pdf> (accessed December 17, 2012).

"Information Note on Matters before the Seventh Session of the Multilateral High-Level Conference on the Conservation and Management of Highly Migratory Fish Stocks in the Western and Central Pacific," Annex 4 of the *Report of MHLC7*.

"Information Note on Matters before the Sixth Session of the Multilateral High-Level Conference on the Conservation and Management of Highly Migratory Fish Stocks in the Western and Central Pacific," *Report of MHLC6*, 37.

"International Plan of Action for Reducing Incidental Catch of Seabirds in Longline Fisheries," Twenty-Third session of COFI, February 1999, <ftp://ftp.fao.org/docrep/fao/006/x3170e/X3170E00.pdf> (accessed August 22, 2010).

"International Plan of Action for the Conservation and Management of Sharks," Twenty-Third session of COFI, February 1999, <ftp://ftp.fao.org/docrep/fao/006/x3170e/X3170E00.pdf> (accessed August 22, 2010).

"International Plan of Action for the Management of Fishing Capacity," Twenty-Third session of COFI, February 1999, <ftp://ftp.fao.org/docrep/fao/006/x3170e/X3170E00.pdf> (accessed August 22, 2010).

"International Plan of Action to Prevent, Deter and Eliminate Illegal, Unreported and Unregulated Fishing," COFI, 2 March 2001, <ftp://ftp.fao.org/docrep/fao/012/y1224e/y1224e00.pdf> (accessed August 22, 2010).

"IOTC's Rules of Procedure," <http://www.iotc.org/English/info/rules_proced.php> (accessed September 03, 2012).

"Large-scale Pelagic Drift-net Fishing and its Impact on the Living Marine Resources of the World's Oceans and Seas," *A/RES/46/215*, 1991, <http://daccess-dds-ny.un.org/doc/RESOLUTION/GEN/NR0/583/03/IMG/NR058303.pdf?OpenElement> (accessed May 5, 2011).

Lotus Case, in *Publications of the Permanent Court of International Justice*, Series A., no. 10, <http://www.icj-cij.org/pcij/series-a.php?p1=9&p2=1> (accessed November 04, 2011).

"Negotiating Text," *A/CONF. 164/13*, Second Session of the United Nations Conference on Straddling Fish Stocks and Highly Migratory Fish Stocks, 1993, <http://daccess-dds-ny.un.org/doc/UNDOC/GEN/N93/702/41/IMG/N9370241.pdf?OpenElement> (accessed February 15, 2011).

"Opening Statement by Chinese Taipei," *Report of the 6th Multilateral High-Level Conference on the Conservation and Management of Highly Migratory Fish Stocks in the Western and Central Pacific*, April 2000.

"Opening Statement by the People's Republic of China," *Report of the 7th Multilateral High-Level Conference on the Conservation and Management of Highly Migratory Fish Stocks in the Western and Central Pacific*, 30 August–5 September 2000, <http://www.spc.int/coastfish/Asides/Conventions/MHLC/MHLC7rep.pdf> (accessed November 09, 2010).

"Opening Statement by Taiwan," *Report of the Seventh Annual Meeting of CCSBT*, 1999, <http://www.ccsbt.org/docs/pdf/meeting_reports/ccsbt_7/report_of_ccsbt7.pdf> (accessed December 15, 2010).

"Opening Statement by Taiwan," *Report of the Sixth Annual Meeting (First Part) of CCSBT*, 1999.

"Opening Statement by Taiwan," *Report of the Fifth Annual Meeting (First Part) of CCSBT*, 1999.

"Participation of Taiwan in the Work of the Inter-American Tropical Tuna Commission," *Resolution 98-09* in IATTC, 1998, <http://www.ofdc.org.tw/fishserv/File/Rule/IATTC_Resolutions_E/98-09.pdf> (accessed June 15, 2012).

"Proposed Amendments to the IOTC Agreement," 2006, <http://www.iotc.org/files/proceedings/2006/s/IOTC-2006-SS3-R[EN].pdf> (accessed September 05, 2012).

"Proposed Amendments to the IOTC Rules of Procedure," 2006, <http://www.iotc.org/files/proceedings/2006/s/IOTC-2006-SS3-R[EN].pdf> (accessed September 05, 2012).

"Proposal for Amendment of the Scheme of Joint International Enforcement of the Fishery Regulations in Regulatory Area," 7 June 1979, *NAFO Annual Report*, vol. 1 (1979), 70–71.

"Protocol Amending the International Convention for the High Seas Fisheries of the North Pacific Ocean," 25 April 1978, <http://www.npafc.org/new/inpfc/INPFC%20_convention.pdf> (accessed January 03, 2012).

"Recommendation by ICCAT to Establish a Multi-Annual Recovery Plan for Bluefin Tuna in the Eastern Atlantic and Mediterranean," Fifteenth Special Meeting of the ICCAT, 2006, <http://www.iccat.int/Documents/BienRep/REP_EN_06-07_I_1.pdf> (accessed December 17, 2012).

"Recommendation on Criteria for Attaining the Status of Cooperating non-Contracting Party, Entity or Fishing Entity in ICCAT," *Report of the Eighteenth Regular Meeting of the ICCAT*, <http://www.iccat.int/Documents/BienRep/REP_EN_02-03_II_1.pdf> (accessed August 27, 2012).

"Record of Decision, Adoption of the Antigua Convention," *Report of the 70th Meeting in IATTC*, 26–28 June 2003, <http://www.iattc.org/PDFFiles2/Antigua%20Convention%20-%20Record%20of%20decision.pdf> (accessed November 10, 2010).

"Record of Discussions between the Commission and Representatives of Taiwan," *Report of the Fifth Annual Meeting (Second Part) of CCSBT*, 1999.

Reparations for Injuries Suffered in the Service of the United Nations, Advisory Opinion: I.C.J. Reports, 1949.

Report of the 12th Special Meeting of ICCAT, 2000–2001, <http://www.iccat.int/Documents/BienRep/REP_EN_00-01_I_1.pdf> (accessed August 27, 2012).

Report of the 15th Regular Meeting of ICCAT, 1996–1997, <http://www.iccat.int/Documents/BienRep/REP_EN_96-97_II_1.pdf> (accessed December 21, 2010).

Report of the 17th Regular Meeting of ICCAT, 2000–2001, <http://www.iccat.int/Documents/BienRep/REP_EN_00-01_II_1.pdf> (accessed December 29, 2010).

Report of the 18th Regular Meeting of ICCAT, 2002–2003, <http://www.iccat.int/Documents/BienRep/REP_EN_02-03_II_1.pdf > (accessed December 29, 2010).

Report of the Eighth Meeting of IOTC, 2003, <http://www.iotc.org/files/proceedings/2003/s/IOTC-2003-S08-R[EN].pdf> (accessed January 11, 2011).

Report of the Ninth Special Meeting of ICCAT, 1994–1995, <http://www.iccat.int/Documents/BienRep/REP_EN_94-95_I_1.pdf> (accessed August 23, 2012).

Report of the First Meeting of the Interim Scientific Committee for Tuna and Tuna-like Species in the North Pacific, 7–10 May 1996, <http://isc.ac.affrc.go.jp/pdf/ISC1pdf/isc1P_rep.pdf>.

Report of the Plenary Session of the Fifth Meeting of the Interim Scientific Committee for Tuna and Tuna-like Species in the North Pacific, 28–30 March 2005, <http://isc.ac.affrc.go.jp/pdf/ISC5pdf/Report_ISC5_Plenary.pdf> (accessed October 07, 2010).

Report of the Second Meeting of the Interim Scientific Committee for Tuna and Tuna-like Species in the North Pacific, 20–23 January 1999, <http://isc.ac.affrc.go.jp/pdf/ISC2pdf/isc2P_rep.pdf> (accessed June 12, 2012).

Report of the Third Meeting of the Interim Scientific Committee for Tuna and Tuna-like Species in the North Pacific, 28–30 January 2002, <http://isc.ac.affrc.go.jp/pdf/ISC3pdf/isc3P_rep.pdf> (accessed June 12, 2012).

Representatives of Belize, Burkina Faso, Chad, Dominica, the Dominican Republic, El Salvador, the Gambia, Grenada, Malawi, Nicaragua, Palau, Sao Tome and Principe, Solomon Islands, Swaziland and Tuvalu to the United Nations Addressed to the Secretary-General, 5 August 2003, *Request for the Inclusion of a Supplementary Item in the Agenda of the 58th Session—Question of the Representation of the Republic of China (Taiwan) in the United Nations. UN Doc. A/58/197*, 2003.

"Resolution 2758 (XXVI)," *A/RES/2758(XXVI)*, <http://daccess-dds-ny.un.org/doc/RESOLUTION/GEN/NR0/327/74/IMG/NR032774.pdf?OpenElement> (accessed June 07, 2012).

"Resolution 99/04 on the Status of Cooperating Non-Contracting Parties," Fourth Meeting of the IOTC, 1999, <http://www.iotc.org/files/proceedings/misc/ComReportsTexts/resolutions_E.pdf> (accessed November 11, 2011).

"Resolution by the ICCAT on Becoming a Cooperating Party, Entity or Fishing Entity," *Report of the 17th Regular Meeting of the ICCAT*, <http://www.iccat.int/Documents/BienRep/REP_EN_00-01_II_1.pdf> (accessed August 27, 2012).

"Resolution on the Establishment of a Working Group to Review the IATTC Convention," *Resolution 98-02* in IATTC, June 1998, <http://www.iattc.org/PDFFiles/C-98-

02%20Convention%20WG%20resolution%20Jun%2098.pdf> (accessed November 10, 2010).

"Resolution on the Participation of a Fishing Entity in the Antigua Convention," *Resolution C-03-09*, in *Report of the 70th Meeting in IATTC*, 26–28 June 2003, <http://www.iattc.org/PDFFiles2/Resolutions/C-03-09%20Participation%20of%20fishing%20entity.pdf> (accessed November 10, 2010).

"Resolution to Establish an Extended Commission and an Extended Scientific Committee," Seventh Annual Meeting 18–21 April 2001 and revised at the Tenth Annual Meeting 7–10 October 2003.

"Revised Draft Convention on the Conservation and Management of Highly Migratory Fish Stocks in the Western and Central Pacific Ocean," *MHLC/WP.1/Rev.4*, 1999, <http://www.spc.int/coastfish/Asides/conventions/MHLC/mhlc_sep99.htm> (accessed November 1, 2010).

"Revised Draft Procedures for Boarding and Inspection Pursuant to the WCPF Convention," 5th session of the WCPFC Preparatory Conference, 2003, <http://www.wcpfc.int/doc/wcpfcprepconwp14-rev-1/draft-procedures-boarding-and-inspection-pursuant-convention> (accessed August 10, 2012).

"Revised Negotiating Text," *A/CONF. 164/13/Rev.1*, Third session of the United Nations Conference on Straddling Fish Stocks and Highly Migratory Fish Stocks, 1994.

"Rio Declaration," United Nations Conference on Environment & Development, 3–14 June 1992, <http://www.unep.org/Documents.Multilingual/Default.asp?DocumentID=78&ArticleID=1163> (accessed February 15, 2011).

"The Second Report on the Regime of the High Seas," *U.N. Doc. no. A/CN.4/42*, 1951, <http://untreaty.un.org/ilc/documentation/english/a_cn4_42.pdf > (accessed November 11, 2011).

"Scheme of Joint International Enforcement of the Fishery Regulations in the Convention Area." *ICNAF Annual Proceedings* 20 (1969–1970): 20–22.

"Statement by Beijing Government concerning *UN Doc. A/58/197*," Website of the Permanent Mission of the People's Republic of China to the UN, 2003, <http://www.china-un.org/eng/lhghyywj/smhwj/wangnian/fy03/t29409.htm> (accessed March 26, 2014).

"Statement by the Fishing Entity of Taiwan to the Admission," *Report of the Ninth Annual Meeting of the Commission*, 2002,

<http://www.ccsbt.org/docs/pdf/meeting_reports/ccsbt_9/report_of_ccsbt9.pdf> (accessed December 19, 2010).

"Statement Made by Taiwan in the Plenary," *Report of the Eighth Annual Meeting of the Commission*, 2001, <http://www.ccsbt.org/docs/pdf/meeting_reports/ccsbt_8/report_of_ccsbt8.pdf> (accessed August 19, 2012).

"Statement Made by the Chairman of the Conference at the Closing of the Second Session," *A/CONF. 164/15*, 1993, <http://daccess-dds-ny.un.org/doc/UNDOC/GEN/N93/443/82/PDF/N9344382.pdf?OpenElement> (accessed February 15, 2011).

"Statement Made by the Chairman of the Conference at the Closing of the Fourth Session," *A/CONF. 164/24*, on 26 August 1994, <http://daccess-dds-ny.un.org/doc/UNDOC/GEN/N94/354/13/PDF/N9435413.pdf?OpenElement> (accessed February 17, 2011).

"Statement Made by the Chairman of the Conference at the Opening of the Organizational Session," *A/CONF. 164/7*, 1993, *United Nations Conference on Straddling Fish Stocks and Highly Migratory Fish Stocks—Selected Documents,* edited by Jean-Pierre Lévy and Gunnar G. Schram, 40. Hague: Kluwer Law International, 1996.

"Summary Record of the Eighth Session of Committee I in the Conference of the Parties of the 15th Meeting of the Convention on International Trade in Endangered Species of Wild Fauna and Flora," *CoP15 Com. I Rec. 8 (Rev. 1)*, 13–25 March 2010.

"Taiwan's Opening Statement," *Report of the Ninth Annual Meeting of the Commission*, 2002, <http://www.ccsbt.org/docs/pdf/meeting_reports/ccsbt_9/report_of_ccsbt9.pdf> (accessed December 19, 2010).

"United Nations Conference on Straddling Fish Stocks and Highly Migratory Fish Stocks," *A/RES/47/192*, 93rd Plenary Meeting, 22 December 1992, <http://www.un.org/documents/ga/res/47/a47r192.htm> (accessed February 28, 2011).

"Western and Central Pacific Fisheries Commission Boarding and Inspection Procedures," 11–15 December 2006, <http://www.wcpfc.int/doc/cmm-2006-08/western-and-central-pacific-fisheries-commission-boarding-and-inspection-procedures> (accessed January 05, 2012).

Non-international Documents/Statutes

"Anti-Secession Law," 14 March 2005, <http://english.peopledaily.com.cn/200503/14/eng20050314_176746.html> (accessed June 01, 2012).

"Law on the Exclusive Economic Zone and the Continental Shelf of the Republic of China," 21 January 1998, <http://law.moj.gov.tw/Eng/LawClass/LawAll.aspx?PCode=A0000010> (accessed April 02, 2014).

"Law on the Territorial Sea and the Contiguous Zone of the Republic of China," 21 January 1998), <http://law.moj.gov.tw/Eng/LawClass/LawAll.aspx?PCode=A0000009> (accessed April 02, 2014).

"Regulations for Coast Guard Administration to Conduct WCPFC High Seas Boarding and Inspection Procedures, 1 May 2010, <http://www.cga.gov.tw/wralawgip/cp.jsp?displayLaw=true&lawId=8a8181d92b5668a7012b5c2785200005&printPage=true> (accessed July 06, 2012).

"Senate Treaty Document 104-24," 2nd Session of the 104th Congress in United States, <http://www.gpo.gov/fdsys/pkg/CDOC-104tdoc24/pdf/CDOC-104tdoc24.pdf> (accessed June 11, 2012).

International Organizations and Similar Bodies

APEC <http://www.apec.org/>
CCSBT <http://www.ccsbt.org/site/index.php>
CITES <http://www.cites.org/>
FAO <http://www.fao.org/>
FFA <http://www.ffa.int/about>
IATTC <http://www.iattc.org/HomeENG.htm>
ICA <http://www.ica-it.org/>
ICN <http://www.internationalcompetitionnetwork.org/>
INPFC <http://www.npafc.org/new/ipnfc.html>
IOTC <http://www.iotc.org/English/index.php>
ISC <http://isc.ac.affrc.go.jp/>
NPAFC <http://www.npafc.org/new/index.html>
SEAFO <http://www.seafo.org/welcome.htm>
UN <http://www.un.org/en/>
WCPFC <http://www.wcpfc.int/>
WTO <http://www.wto.org/>

INDEX